PREFACE

In presenting such a radical treatment of mechanical drafting, both as to arrangement and selection of material, a few words of explanation may not be amiss. The author is decidedly opposed to giving in a college course in mechanical drafting any work of a purely abstract geometrical nature. Most students have gotten clear conceptions of the geometrical figures in the high school, and to attempt to teach the use of instruments in the drafting of such figures is a gross waste of the student's time in the already too short college course. Long continued trials have definitely proven that the student can just as well, and perhaps better, be taught the use of instruments on work that will at the same time have intellectual value.

In the usual text on drawing the subject is presented in the old order of logically progressive chapters; one chapter on one subject, another on a different subject, and so on. This has always made it extremely difficult to design a satisfactory course of drafting problems in which the instructor has not had to do an excessive amount of lecturing. If a student is to get a thing and keep it, he must be made to dig it out himself, within reasonable limitations. Lecturing simply deducts so much from the time when the student is self-active.

With these ideas in mind the author has planned a very flexible course and written a text to suit the course. The course is divided into distinctly logical steps, and each successive step is made a separate block of work to be finished before the next is begun. All of the information needed by the student in performing the work of block No. 1 is compiled into Lesson 1, and no more. All additional information needed for block No. 2 is given in Lesson 2, and so on. Illustrations are included for everything that can be illustrated, for the student's *visual* memory seems to be better than his *word* memory.

In using this set of lessons the author has discontinued lecturing entirely. When a class is ready for the next block of work, the corresponding lesson is assigned; the students are required to recite on the subject matter, quizzes are given, and freehand sketches on the blackboard illustrating all points are called for. Successive trials have shown that the students thus retain about twice as much knowledge as before under the lecturing plan.

A glance thru the text will show that it has been made as practical as possible. Drafting is not so much a theoretical subject; it is intended to be used and should be practical. Furthermore, the author has kept in mind the fact that this is an age of intense competition, and that a draftsman should never draw a piece of any machine until it has been decided on what machine or by what means each surface can be most rapidly and cheaply finished. If a certain surface can be machined more cheaply on a milling machine than on a shaper or planer, then bosses and ribs should not be so placed as to prevent milling machine work. To assist in giving the student this practical training, illustrations of all the standard machines are given, with tools and accessories, so that he may gradually learn to couple his drafting with shop operations.

H. W. MILLER.

Urbana, Illinois, July, 1912.

2

CONTENTS

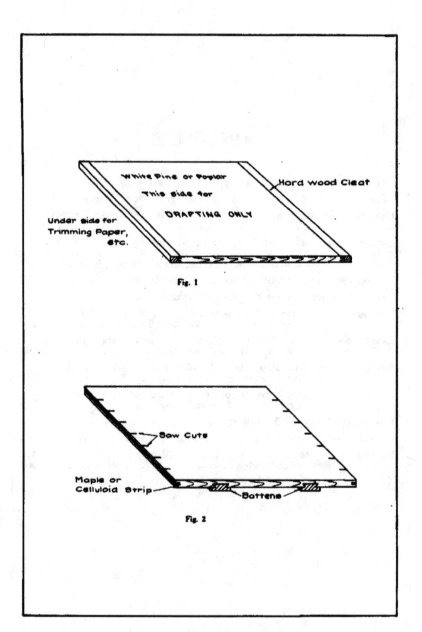

White Pine or Poplar
This side for
DRAFTING ONLY

Hard wood Cleat

Under side for
Trimming Paper,
etc.

Fig. 1

Saw Cuts

Maple or
Celluloid Strip

Battens

Fig. 2

4

MECHANICAL DRAFTING

LESSON 1

USE OF INSTRUMENTS

DRAWING-BOARD

(1) **Construction.** Drawing-boards are made of either poplar or white pine, the right and left edges, Fig. 1, being reinforced by cleats of some harder wood. These cleats serve both as stiffeners and as runners for the easy sliding of the T-square. The better grades of small boards are reinforced on the back by two *battens*, Fig. 2, and ordinarily, have inserted in their right and left edges a wearing strip of either hard maple or celluloid, instead of the cleats of Fig. 1. It will be noticed in Fig. 2 that the right and left edges of the second class of board are broken at intervals by saw cuts which prevent the inserted strip of hard wood from expanding and splitting the wood.

Use. The two sides of the type of board shown in Fig. 1 have very definite uses if the board is to be kept in shape for good drafting. The one side for *drafting only,* the other for any necessary rough work, *trimming paper,* etc. **Never trim paper on the drafting side.**

PAPER

(2) **Quality.** A novice cannot obtain good work from poor material, hence it is imperative that the beginner in drawing use the best quality of paper obtainable. A heavy, hard surface paper of the quality of Keuffel & Esser's **Normal,** or E. Dietzgen's **Napoleon** is recommended.

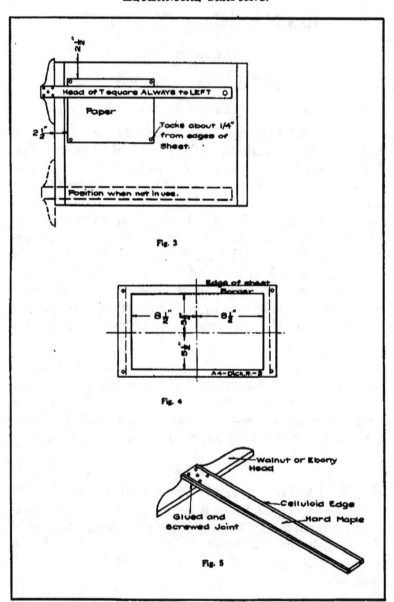

Fig. 3

Fig. 4

Fig. 5

Position of paper on board. In tacking the paper to the board, Fig. 3, keep the sheet well toward the *top* and to the *left;* about two and one-half inches from the upper and left edges. This should be done in order that the draftsman may work to advantage on the bottom of the sheet, and that it may not be necessary to work to any great extent on the end of the T-square blade, which cannot be prevented from springing slightly.

Tacking sheet. Place *upper left* corner of sheet in approximately correct position and tack to the board, Fig. 3, placing tacks close to the corners of the paper. Then after lining up the upper edge with the upper edge of the T-square blade, stretch sheet and tack *upper right* corner. The lower edges may be tacked down in any order; or, after some experience, may be left untacked, as these tacks have a tendency to interfere with the T-square and triangles.

BORDER LINES

(3) The *border lines* as well as all other construction work done by the draftsman should be placed in by measurements from *center lines* and **not** from the *edges* of the *sheet*. In the case of the border lines, the measurements are made from the horizontal and vertical center lines of the sheet, Fig. 4.

Dimensions. The dimension for the border lines on all of the work in this course will be 11"x17". Size of sheet when finished, 12"x18". See Fig. 4.

T-SQUARE

(4) **Construction.** Both the **blade** and the **head** of the T-square, Fig. 5, are of hard wood, hence the glue

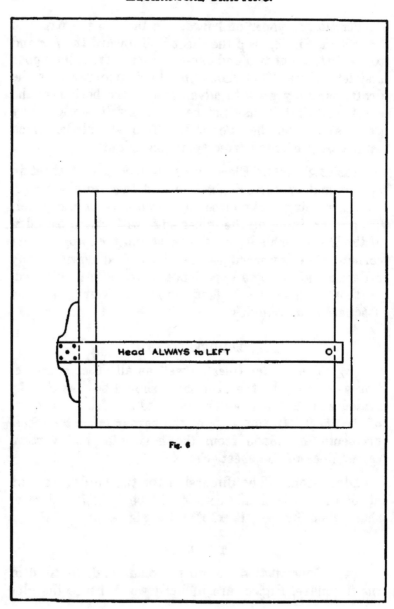

Fig. 6

cannot cement them very tightly together; neither do the
short screws hold very firmly; a fall, even to the floor,
may break the joint and damage the T-square. Keep the
T-square out of danger of any such fall.

In case the joint breaks, take out screws, rough both
head and blade with coarse sandpaper, coat well with
Lepage's glue, place blade at 90° with head with triangle,
tighten screws and let stand a day. Then take out screws
and put in round-headed wood screws long enough to run
through and project an eighth of an inch or more. File
off screws carefully. Be sure screws are tight. It may
be advisable to bore small holes entirely through head
and blade for these large screws, to prevent splitting of
the wood.

Position on board. In drafting, a right-handed man
should keep the head of the T-square to the *left,* Fig. 6,
in order that he may handle it with his left hand, leaving
the right free for drafting. **Never place the T-square in
any other position on the board,** as the edges of the board
seldom form a rectangle nor is the head of the T-square
likely to make exactly 90 degrees with the blade.

Use of blade. The *upper edge* of the blade should be
used for *drafting* only; and the draftsman will do well
to take excellent care of this edge, for once niched or
dented the instrument is practically ruined for good work.
The *lower edge* may be used as a *cutting ruler* but **never**
for *drafting.*

Position when not in use. Any draftsman profits by
keeping his drawing instruments in certain definite
places, so that as far as possible he may keep his attention entirely on his work, handling his instruments sub-
consciously. It is found most convenient to slip the

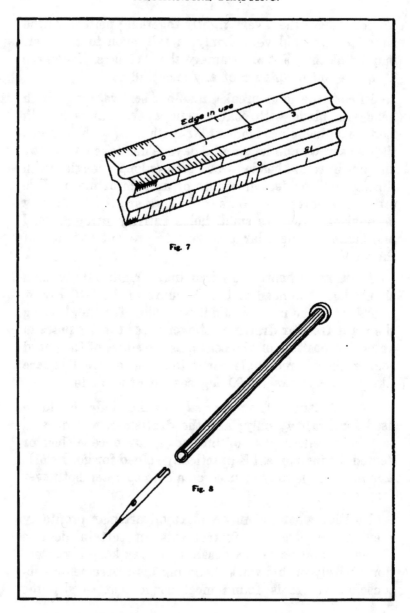

Fig. 7

Fig. 8

T-square to the bottom of the board, when not in use; it is here out of danger, out of the way, yet easily accessible.

To keep clean. A drawing may easily be smudged by a dirty T-square, so it will be well to give the blade a thoro cleaning with a damp cloth or piece of art gum at frequent intervals.

SCALE

(5) **Care.** The scale should **never** be used as a *ruler* because, as a drafting instrument its efficiency depends upon the condition of its edges, and these edges can easily be defaced by misuse and the instrument badly damaged. Furthermore, the boxwood of which the scale is made warps quite easily; hence, the edges of a triangular scale will seldom be found perfectly straight.

Use. On inspection, Fig. 7, it is seen that the numerals are all placed on the scale so as to appear upright only when one works *over the top* of the scale or on the edge *away* from the draftsman and *not toward* him. All dimensions should be taken directly from the scale as it lies on the drawing and **not** by means of the dividers. The *needle point* is the best aid in obtaining dimensions with perfect accuracy. The pencil point is a poor substitute for the needle point.

NEEDLE POINT

(6) **From Richter Set.** An excellent *needle point* for obtaining dimensions may be made up by inserting into the long *knurled barrel* furnished with every set of Richter instruments, the *small point* which is provided for converting the large compass into a set of dividers, Fig. 8.

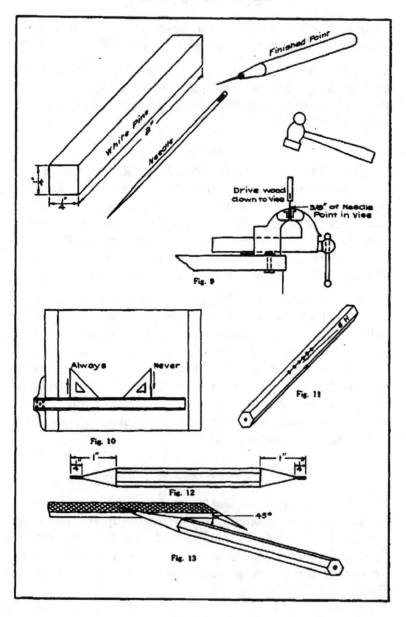

Fig. 9

Fig. 10

Fig. 11

Fig. 12

Fig. 13

To make in shop. A *needle point* may be easily made from a strip of white pine ¼"x¼"x2" and a medium size sewing needle.

Construction: Fig. 9. Insert the needle in a vise, point *down*, with about ⅜" of the point in the vise, and carefully drive the strip of pine over the exposed part of the needle; the wood may then be shaved round and pointed slightly at the needle end.

TRIANGLES

(7) In drawing *vertical lines* with triangles the *vertical edge* should always be to the **left** or *toward* the *head* of the T-square, Fig. 10.

PENCILS

(8) **Numbering.** Before sharpening either end of the pencil, cut with a pen knife a number of nicks toward the center to correspond to the degree of hardness; e. g., four nicks for 4H, six for 6H, etc., Fig. 11.

Sharpening wood. In sharpening, trim the wood carefully on both ends of the pencil, back a distance of about one inch from the ends, leaving about ¼" of the lead exposed, Fig. 12. One end is to be sharpened to a *round point*, the other to a *wedge*. In shaping up both of these points, use the pencil point file provided in the kit.

Round point. In shaping up the round point, hold the pencil at an angle of about 45 degrees with the axis of the file. As the lead travels over the file, Fig. 13, revolve the pencil slowly between the thumb and fingers, attempting to give it a complete revolution with each stroke. The lead may thus be easily sharpened to a perfect cone. In this sharpening be careful that the point extends the *full length* of the lead exposed.

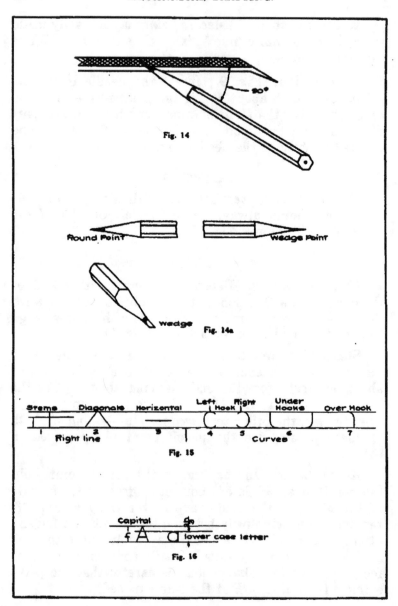

Fig. 14

Round Point　　　　　　　　　Wedge Point

wedge　　Fig. 14a

Stems　　Diagonals　　Horizontal　　Left Hook　Right　　Under Hooks　　Over Hook
1　　　2　　　　3　　　4　　5　　　6　　　7
Right line　　　　　　　　　　　Curves

Fig. 15

Capital　　up
f　　　　lower case letter
Fig. 16

Wedge point. In sharpening the *wedge point* hold the pencil perpendicular to the axis of the file, Fig. 14, and so inclined to the plane of the file that the lead may be sharpened the full quarter inch exposed.

Use of points. The *round point* should be used for drawing *short lines* and *lettering,* the *wedge point* for *long lines;* the round point dulls rapidly in drawing a long line and will make a line of varying weight.

ERASERS

(9) If an eraser becomes apparently *greasy* and smudges instead of cleans a drawing, it may easily be cleaned by rubbing it with another eraser or by rubbing it on the rough surface of the drawing-board itself.

OFFHAND LETTERING

(10) Offhand letters, tho apparently complex in their construction, when analyzed into their component parts, are found to be composed of just seven elements, Fig. 15; three of these are *right lines,* and four, comparatively *simple curves.*

Height of letters. From the printer's custom of keeping the capital and small letters in different parts of his type case they have quite generally become known as **capitals,** and **lower case** letters, Fig. 16, rather than **small** letters. Letters may be of any height; however, in every case the *height* of the *lower case* letters, Fig. 16, is two-**thirds** of the *height* of the *upper case* or *capital* letters. Dimensions for letters are hence given as 3/16″x1/8″, 1/8″x1/12″, etc., the first dimension giving *h,* the *height* of the *capital,* the second 2/3h, the *height* of the *lower case* letters.

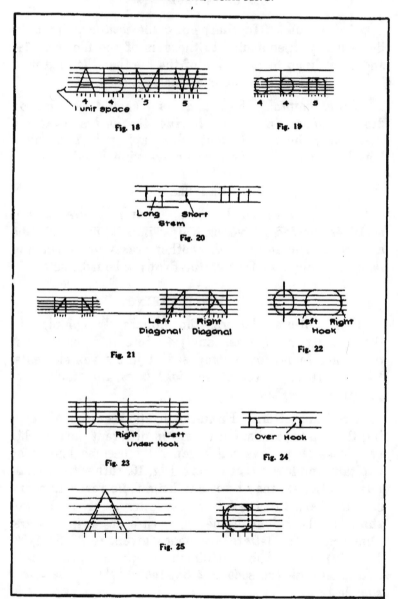

Fig. 18

Fig. 19

Fig. 20

Fig. 21

Fig. 22

Fig. 23

Fig. 24

Fig. 25

Width of letters. If the *height* of the letter, for example the *capital,* be divided into *five equal parts,* one of these five equal parts is known as a **unit space,** Fig. 18. If then a rectangle be constructed with a height equal to the height of the letters and a width of four *unit spaces,* it will enclose each of the letters of the alphabet except E, F, P, R, M and W; that is, all of the letters of the alphabet except E, F, P, R, M and W are *four unit spaces wide,* E, F, P and R *three and one-half units,* and M and W *five.* The same is true of the lower case alphabet; it must be understood, however, that in the lower case alphabet, the *width* of the letter will be four-fifths of the *height* of the *lower part* of the letter, Fig. 19. The length of the long stems will have no effect on the width of the letters.

Distorted letters. Letters (except E, F, P, R, M and W) whose *width,* as has just been explained, is *four-fifths* of the *height* are known as **normal** letters; if the *width* of the letter be *less* than four unit spaces, it becomes a *distorted* letter and is known as a **compressed** letter; if on the other hand the *width* is *greater* than four unit spaces it becomes an **expanded** letter.

COMPONENTS OF LETTERS

(11) **Stems.** The *vertical components* of letters are known as **stems**; if they have a *length* equal to the *height* of the *capitals* they are known as **long stems,** Fig. 20, and enter into the composition of such letters: *B, D, T, b, h, p, q,* etc. If the *stem* has a *length* equal to the *height* of the *lower case* letters it is known as **a short stem;** such stems of course enter into the composition of only the lower case letters. In making either of these stems with either the pen or pencil, the direction of the stroke is *down.*

Left diagonals. If in a rectangle with a *height* equal to the height of either the *capital or lower case* letters and width equal to four-fifths of this height, Fig. 21, a *diagonal* be drawn from the *upper right* to the *lower left* corner, we have what is known as a **left diagonal,** from the direction of the stroke in formation. Such diagonals enter into the composition of the letters *x, z, v, y, M, W, V,* etc.

Right diagonals. The *diagonal* from the *upper left* to the *lower right* corners of the rectangle just mentioned, Fig. 21, is known as a **right diagonal,** from the general direction of the stroke in formation; this diagonal enters into the composition of letters *x, y, k,* etc., also *N, M, W.* etc.

Horizontals. The *horizontal* components of such letters as H, L, E, F, etc., are known as **horizontals;** the direction of the stroke in formation is from *left* to *right.*

Left hooks. If the letter O be cut in half by a vertical line, Fig. 22, the *left half* is known as a **left hook,** the direction of the stroke being from the *top toward* the *left, down,* then *toward* the *right;* this component, with slight variations, enters into the formation of the lower case letters *a, c, d, e, g, o, q* and *s.*

Right hooks. The *right half* of the letter *O,* Fig. 22, is known as a **right hook,** the direction of the stroke being from the *top toward* the *right, down,* then *to* the *left.* This component enters into the formation of the capital letters *D, P, O, Q, R* and *S,* and lower case letters *b, p, o* and *s.* Tho apparently no more difficult to form than the left hook, this right hook seems to present the main difficulties of the alphabet; for the letters *b, p* and *s* are the most difficult in the whole alphabet to form properly and in this group the letter *b* seems to be the most difficult; tho apparently identical in construction with the letter *p, b* is in fact the bugbear of every man learning to letter and should receive the most practice.

Under hooks. If the capital letter *U* be cut in half by a vertical line, Fig. 23, the *left half* is known as a **right under hook,** the direction of the stroke being *down* and *to* the *right*. The *right half* of the letter *U* is known as a **left under hook,** the direction of the stroke being *down* and *to* the *left*. These two components do not present any great difficulty to the average draftsman and the letters *u, g, j* and *y, J* and *U,* of which they are the chief components, will not need a great amount of practice.

Over hook. The *curved* component of the lower case letter *h* is known as an **over hook,** Fig. 24, the direction of the stroke being *to* the *right* and *down;* this component enters into the formation of the letters *h, m, n* and *r.*

(12) **The most difficult letters.** The most difficult letters to form perfectly are found in the lower case alphabet and are, in the order of their difficulty, *b, p, y, s, r* and *g.* As has been mentioned before the letter *b* seems far more difficult to form than the letter *p* and should receive the bulk of the extra practice.

Easy style of letter. To the beginner, the *normal* style of letter, with a width of four-fifths of the height, seems to be extremely difficult to form, while the *expanded* style, Fig. 25, has a number of distinct advantages; perhaps the fact that horizontal lines prevail in actual life makes these expanded letters easier of formation, the *horizontal* predominating in this style. Furthermore, *expanded* letters have a most excellent appearance irre-

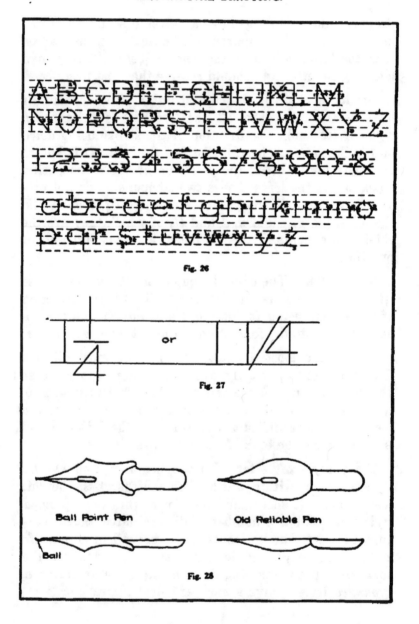

ABCDEF GHIJKLM
NOPQRSTUVWXYZ
1234567890&
abcdefghijklmno
pqrstuvwxyz

Fig. 26

1/4 or 1/4

Fig. 27

Ball Point Pen   Old Reliable Pen

Ball

Fig. 28

spective of the extent to which they have been expanded; that is, no matter how great the distortion may be, the effect is invariably good; for these reasons the beginner will find it well to adopt this *expanded* style; at least until the formation of the letters has become a habit.

NUMERALS

(13) **Height.** When *numerals* are used with letters they are given a *height* equal to the *height* of the *capitals*, Fig. 26.

Fractions. Theoretically the height of the numerals of a fraction should not be as great as the height of integral numerals; however, dimensions do not look at all out of proportion when all of the numerals, whether fractional or integral, are given the same height. Guide lines should be ruled for the integral numerals, Fig. 27; however, they need not be ruled for the fractions, as the heights of these can easily be approximated.

LETTERING PENS

(14) **Styles.** The styles of pens that have been found best for lettering are shown in Fig. 28. The **ball point** makes a rather heavy line; however, it has the distinct advantage of making a line of very uniform weight and works well for beginners. With the **old reliable** pen it is possible to make lines of much lighter weight and this pen is usually preferred by draftsmen. To those who have a great amount of lettering to do the writers recommend **Moore's Non-leakable** fountain pen; from severe trials it has been found to work with perfect satisfaction and its convenience cannot be over estimated. A *medium fine* pen makes the most uniform letters.

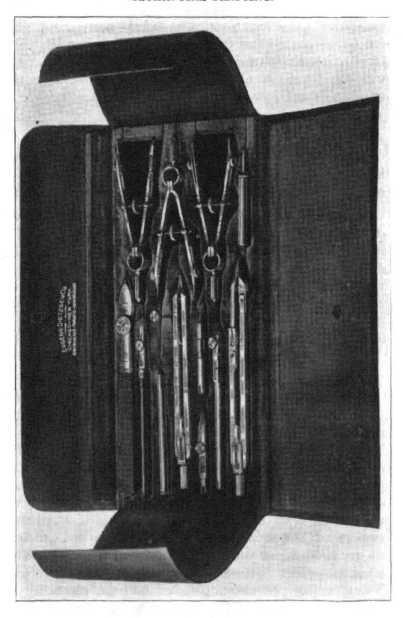

CLEANING PADS AND BLOTTERS

(15) **Chamois roll or block.** Inasmuch as water-proof drawing ink dries so rapidly the pen should be cleaned thoroly with cloth or chamois before each refilling. In addition to this cleaning it will be found possible to obtain more clear cut letters if after each three or four letters the point of the pen is scraped over a piece of chamois. A convenient scraper may be made by rolling up a 2″x4″ piece of chamois and binding it with a rubber band, Fig. 29, or by pasting a 2″x2″ piece on a small block of wood.

Blotters. Never attempt to blot drawing ink. The ink is too heavy to be absorbed by blotting paper. Always permit the ink to dry. A puddle of ink may, however, be removed by the careful use of the corner of a blotter.

Use of Ink Bottle. For proper use of ink bottle see Art. 21 and Fig. 41.

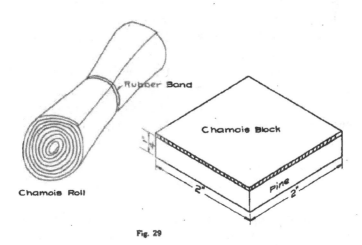

Fig. 29

LESSON 2

USE OF INSTRUMENTS

LARGE DIVIDERS

(16) **Adjustment of the points.** With *Richter* instruments it will always be found possible to adjust the points of the various tools to any desired length; so, before attempting to use the *large dividers* be sure that the points are adjusted to exactly the same length and that they are in perfect shape. In case the points of the *Gem Union* instruments are not of the same length, it will be necessary to grind the long point down on a small **carborundum** stone. Keep points always in perfect shape for good work.

Opening and setting. It is desirable always to handle each instrument with the *right hand* unaided by the *left;* this permits of much more rapid work and the habit is not difficult to acquire. To *open* the divider, *insert* the *thumb between* the *legs,* prying them apart a short distance until the fingers may be inserted and the *one leg* grasped *between* the *first* and *second fingers,* the *other between* the *third finger* and the *thumb;* the head of the instrument should rest against the knuckle of the first finger. Holding the instrument in this position it is found easily possible to adjust the points to any desired distance.

To place point. *To place* the one *point* of the divider at any *point* on the sheet, rest the *wrist* at a convenient distance from the point; it will then be found easily possible to place the *point* of the *leg between* the *third finger* and the *thumb* in any desired position. Raising the wrist and keeping the little finger on the paper, the other leg can now be adjusted for any desired distance. It is perhaps as good practice and may be found easier for some to steady the hand thruout the operation by merely resting the little finger on the paper, instead of the wrist.

Stepping off distances. After the points have been placed as desired, to *step off* a certain distance a number of times, raise the first finger to the top of the head, then, releasing the other leg, *grasp* the *head between* the *first finger* and the *thumb* and step off the distance by swinging the dividers alternately *over* and *under*. Handling the instrument in this way it will not be necessary to take a new grip on the head thruout the operation.

BOW DIVIDERS

(17) **Adjustment.** (See **Adjustment** for **Large Dividers.**)

Placing at center. (See same for **Large Dividers.**)

Opening and closing points. With the *center adjustment* instrument, which is always preferable to the *side adjustment,* after placing the one point at a given point · on the sheet, *raise* the *first finger* to the *head* and *turn* the *adjustment screw between* the *second finger* and *thumb* until the points are apart as desired.

Stepping off distances. (See same for **Large Dividers.**)

TRIANGLES

(18) **To clean.** The surface of the celluloid triangles quickly becomes smudged from erasings and pencil dirt that may be on the drawing; hence, they must be

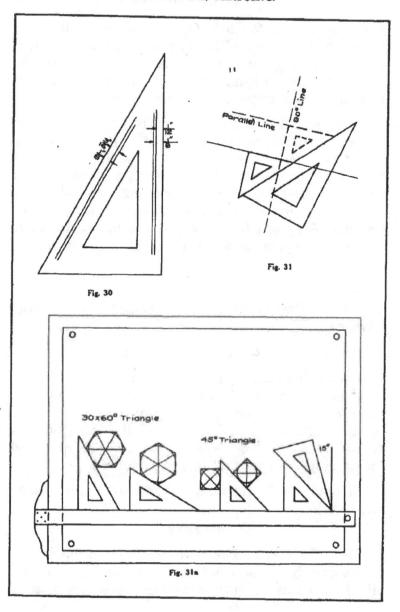

Fig. 30

Fig. 31

Fig. 31a

cleaned **frequently** with soap if the drawings are to be kept in good shape.

Letter guide lines. For the easy ruling of letter guide lines without the use of the scale and needle point, it is suggested that along the edges of the 30x60 *triangle* light lines be scratched with the needle point as follows: Along the *hypothenuse* and 1/8″ from the edge scratch carefully a fine line; also a second line 3/16″ from the edge; along the long leg scratch two lines 1/8″ and 1/12″, respectively, from the edge, Fig. 30. After the lines have been scratched they should be smeared over with India ink, rubbing the ink into the scratches with the fingers. After the ink has dried for a few minutes the surplus may be rubbed off with a cloth. Turning the triangles over with the scratched lines against the paper, it is seen that they now stand out very sharply and may be used in ruling guide lines for any necessary lettering.

Parallel and perpendicular lines. In Fig. 31 are shown a number of methods of obtaining a series of parallel lines, or lines perpendicular to given lines, by means of the *triangles* and *T-square*.

BOW PENCIL

(19) **Hard lead.** To obtain satisfactory work from the *bow pencil* the lead should be extremely hard, at least 6H. Ordinarily, the lead supplied with instruments is not more than 2 or 3H and wears down too rapidly. Try the lead before using it on a drawing and if found soft substitute for it a piece of lead from a 6H pencil.

Sharpening lead. Adjust the lead until it is the *same length* as the *needle point*, then shape up the *wedge point*

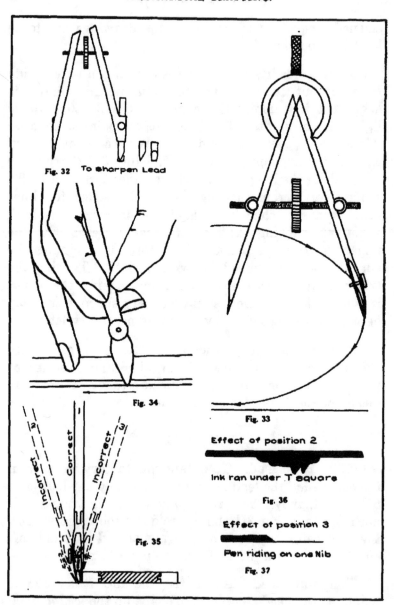

Fig. 32   To sharpen Lead

Fig. 34

Fig. 33

Fig. 35

Effect of position 2

Ink ran under T square

Fig. 36

Effect of position 3

Pen riding on one Nib

Fig. 37

as shown in Fig. 32. Grind the outside bevel at an angle of about 30 degrees until the cut has run about three-fourths across the end of the lead; then tip it off slightly at a similar angle on the inside. The lead, thus sharpened, both wears well and gives most satisfactory work. **Never sharpen the lead of the compass or bow pencil to a round point.**

Adjustment to any radius. In *adjusting* the *points* to any *desired radius,* instead of obtaining the dimension *directly* from the *scale,* it will be better to *transfer* this *radius* to the *paper* by means of the *scale* and *needle point,* and set the bow pencil from this as explained under **Large Dividers.**

Describing arcs. In *describing* an *arc* with the bow pencil, the direction of motion of the lead should be *clockwise* and thru the total length of the desired arc before taking the lead from the paper. See Fig. 33.

RULING PEN

(20) **Manner of holding.** The *ruling pen* should be held *between* the *first* and *second fingers* and *thumb* as shown in Fig. 34. In ruling lines, the *adjusting screw* should be turned *from* the user.

Position of pen. Unless care is taken to keep the pen in a *vertical plane* thru the *edge* of the *T-square blade* or *edge* of the *triangle,* Fig. 35, trouble may be experienced in the ink running under the T-square blade, Fig. 36, or in a badly broken line, Fig. 37.

Tilted in the direction of motion. For best results the pen should be *tilted* slightly in the *direction* of the motion, Fig. 34; this permits one to inspect the work of

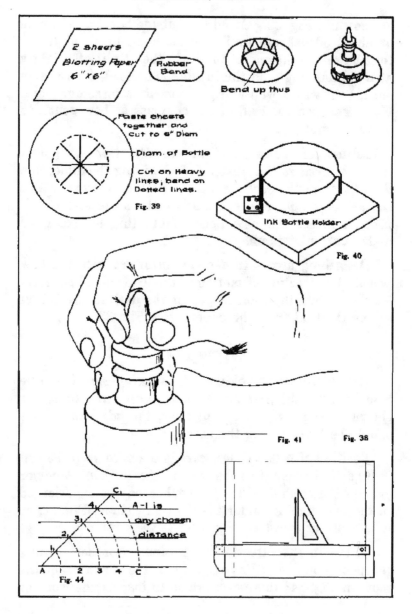

2 Sheets
Blotting Paper
6"x6"

Rubber Band

Bend up thus

Paste sheets together and cut to 6" Diam

Diam. of Bottle

Cut on Heavy lines, bend on Dotted lines.

Fig. 39

Ink Bottle Holder

Fig. 40

Fig. 41

Fig. 38

A-1 is any chosen distance

Fig. 44

the pen as it travels. A greater angle than 10 or 15 degrees may however permit the ink to run down and cause a blot.

To fill pen. To *fill* the pen, always use the *quill* supplied on the stopper of the ink bottle. **Never dip the pen into the ink.** If by chance any ink has gotten on the outside of the pen, wipe it off carefully before using; it may save a serious blot.

Direction of motion in ruling lines. In *ruling lines,* either with the *pen* or *pencil,* the direction of motion should be *from left* to *right* or from *bottom* to *top* of the sheet, Fig. 38. .Ruling lines thus, it is always possible to see what the pen or pencil is doing. **Never** rule lines **down** the sheet unless they are oblique and are being put in with the triangle.

INK BOTTLES

(21) **Holder.** A convenient holder for the ink bottle may be made from two sheets of blotting paper and a rubber band as shown in Fig. 39. A holder of some kind is advisable and one such as this answers a double purpose. A holder of the style shown in Fig. 40 may be purchased from any of the instrument companies.

Closed when not in use. India ink is very heavy and dries quite rapidly; hence, if the stopper is left out of the bottle even for several hours the ink may become so heavy as to make it impossible to obtain good work from it. Be sure to **close** the bottle carefully after each refilling of the pen.

To open bottle and fill pen. To open the bottle without danger of upsetting, *grasp* the *neck between* the *third*

and *little fingers*, Fig. 41, and the *stopper between* the *first finger* and *thumb* of the **same hand**; after removing the stopper, *place* the *quill between* the *nibs* of the *pen* and fill as desired.

MECHANICAL LETTERS

(22) Little need be said in explanation of the following alphabet of mechanical letters, Figs. 42 and 43; this style is given mainly for its simplicity of construction, and tho perhaps slightly defective in minor details, the letters, when combined into words according to the scale in Lesson 3, will have a very symmetrical and uniform appearance.

UNIT SPACE

If the total *height* of the letter be divided into *five equal parts, one* of these five parts is known as a **unit space.** In construction, these unit divisions may be obtained as shown in Fig. 44. For rapid and easy construction of block letters, all of the space lines thru 4_1, 3_1, 2_1 and 1_1 should be ruled in; however, these lines should be very light.

WIDTH OF LETTERS

Inspection of the alphabets, Figs. 42 and 43, shows that each letter of the alphabet except *E, F, P, R, M* and *W* is *four unit spaces wide,* while *E, F, P* and *R* are *three* and *one-half* and *M* and *W five.* The alphabet is hence so simple that its details can easily be memorized.

THICKNESS OF LETTERS

In every case, whether the letter be *normal, compressed,* or *expanded,* the *thickness* may be kept **one unit.**

If space permits and it is desired, any r
chosen for obtaining an exaggerated wid
Fig. 45; however, all *radii* of arcs and
the letter are in terms of the *unit spac*
true of a *compressed* letter.

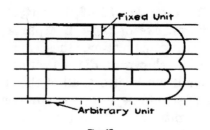

Fig. 45

LESSON 3

NAME PLATES AND TITLES

(23) **Name plates.** A *name plate* for any machine or piece of structural work should contain the following information: *Name* of the *machine* (unless it is so common as to be perfectly familiar to everyone), *name* of the *manufacturing company,* and *address* or *location* of the *company's works* or *factories.*

(24) **Drawing titles.** A *working drawing title* should contain the following information: *Name* of the *piece of machinery* drawn, *name* and *address* of the *manufacturing company, initials* of the *draftsman, checker,* and *tracer, scale, drawing number,* and other necessary *filing data.*

GENERAL ORDER OF WORK IN CONSTRUCTION OF WORKING DRAWING TITLE

(25) **Given data.** In making up a *working drawing title* the draftsman ordinarily has given him a certain amount of data as follows: "Details *of* Horizontal Milling Machine, *manufactured by the* Landis Tool Company, Waynesboro, Penna; drawn by (R. C. S.), checked by (————), traced by (————), scale ½"=1", *drawing finished* April 2nd, 1915." The above material, condensed, must be placed within a given title space, perhaps 3"x5".

Elimination of unnecessary material and arrangement into groups. In order that the given material may be placed within the given title space, every unnecessary

word must be eliminated. Running thru the given data
it is seen that the words italicized can be omitted
without the least danger of misunderstanding the
remainder. With these words omitted the remaining
data seems to group itself naturally as follows:

(1st prom.) **Horizontal Milling Machine.**

Details

(2nd prom.) **Landis Tool Co.**

(3rd prom.) **Waynesboro, Pa.**

Drawn by (——), Checked by (——), Traced by (——),
(4th prom.) **Date (——), Scale (——).**

Order of prominence. In a *drawing title* as well as
in a *name plate*, certain groups of words are more important than others. In the *drawing title* the *name* of the
piece of machinery is, of course, given most prominence;
while, in the case of the *name plate* the *name* of the *manufacturing company* should be given *first* prominence. In
the *drawing title* the *name* of the *manufacturing company* will be given *second* prominence, *address* of the
company *third*, and the remaining information, being
about equally important, should be given *least* prominence and arranged as desired. The word **"Detail"** or
"Assembly", which may be either included or omitted
as desired, will not figure in the order of prominence.

Methods of securing prominence. In both advertis-
ing and drafting there are in use two methods for securing prominence of any one group of words over another.
The one most generally used is **variation** in **height** of
the letters of the various groups, to correspond in general

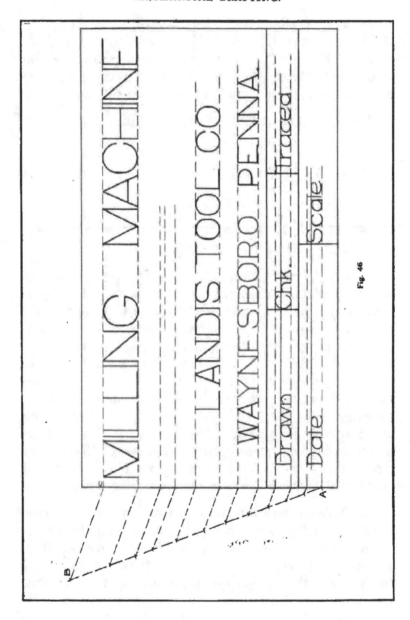

Fig. 46

to the order of prominence established; if this is not possible thru lack of space, a **distorted** letter, either of odd construction or of the *compressed* or *expanded* style, may be used. One may then depend on the odd appearance of the letters to give to that group of words the desired prominence. In the case of *drawing titles* and *name plates* the first method, i.e. *variation in height* of letters is preferable.

Margins and margin lines. Before sketching in the guide lines for the various groups of words, *margin spaces* should be determined and *light margin lines* ruled in Fig. 46. In doing this certain rules of design must be adhered to; the upper and lower margins may be of any desired width; however, for best appearance they must be *equal.* The right and left margins, tho not necessarily equal to the upper and lower, must be *equal* to *each other.* In a rectangular space whose *width* is *greater* than the *height,* better effect is obtained if the *right* and *left margins* are made slightly *greater* than the *upper* and *lower;* if the reverse is true of the rectangle then the right and left margins should be less than the upper and lower.

Guide lines. To obtain the guide lines for the various groups of words, produce the lower margin line to the left until it intersects the border of the title space at A, Fig. 46, and from this point draw up and toward the right a line at an angle of about 60 degrees with the horizontal. Selecting any desired distance as the relative height of the letters of the lower group, lay off this distance from A along the 60 degree line; then a relative distance for the space between this line and the line next above; next a relative space for the letters of the next group and so on until all of the groups have been accounted for along the 60 degree line. From the last

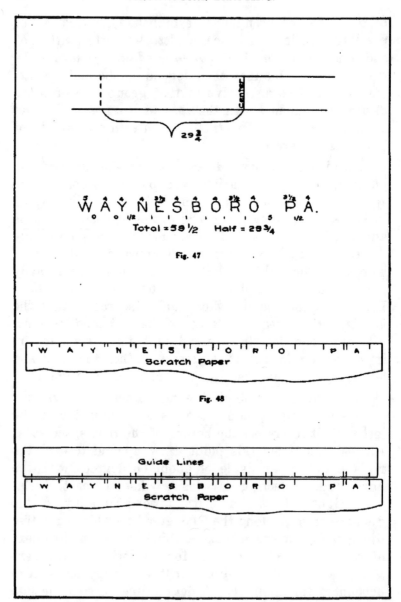

Total = 59 ½   Half = 29 ¾

Fig. 47

Scratch Paper

Fig. 48

Guide Lines

Scratch Paper

point, B, draw a line, B C, as shown in figure and from the various points along the line A B draw lines parallel to B C until they intersect the border line, A C; the required guide lines will then be found the same relative distances apart as the points plotted on the line A B.

Spacing of letters. Before attempting to place in any of the letters the *value* of the *unit spaces* of the various groups of letters must be determined. Tho the various lines of letters may not require all of the horizontal space allotted to them, for best appearance they must be placed centrally; i.e., with equal margins at their right and left. To accomplish this, rule in a vertical center line of the title space and use either of the following methods:

(1) MATHEMATICAL METHOD

Taking for illustration the group "Waynesboro Pa.", sketch roughly the letters of these two words on a piece of scratch paper, spacing liberally, Fig. 47; next place *above* each letter its *width* in *unit spaces* and *between* the various letters the *number* of *spaces* required to *separate* letters, obtaining these from the table, Fig. 49. *Between* the two *words* allow at least *five unit spaces*. The sum of all of these spaces is $59\frac{1}{2}$; one-half of $59\frac{1}{2}$ is $29\frac{3}{4}$; stepping off $29\frac{3}{4}$ spaces to the left from the center gives us the point at which the letter **W** of this line should start.

(2) SCRATCH PAPER METHOD.

Select any point close to the left end of the straight edge of a sheet of scratch paper, Fig. 48, and from this point step off with the large and small dividers the proper number of unit spaces in succession, for the various letters and spaces between letters, marking with the divider points the location of the *beginning* and *end* of each letter. Placing the scratch paper centrally along the lower guide line of the space into which this group of letters is to go, mark with the needle point the position of the beginning and end of each of the letters. This method has the advantage of centrally placing the entire group and of locating the various letters at the same time.

TABLE OF LETTER SPACES

(26) To obtain the *space*, in *units*, to be allowed *between* any letter, e.g. **A**, and any letter of the alphabet which may follow it in a word, it is seen in the table, Fig. 49, that between **A** and any letter of the alphabet except **T, V, W** and **Y**, should be left one-half *unit space*, while between **A** and **T, V, W** or **Y**, *no spacing* should be left. In every case the spacing given is that to be allowed between the letter given in the first column and any letter of the alphabet which may *follow* it in a word.

Fig. 49
Table of Block Letter Spacing

Letter	Regular Spacing	Spacing for Exceptions	Exceptions	Letter	Regular Spacing	Spacing for Exceptions	Exceptions
A	1/2	0	T, V, W, Y	N	1	1/2	A, T, V, W, Y
B	1	1/2	A, T, V, W, Y	O	1	1/2	"
C	1	1/2	"	P	1	1/2	"
D	1	1/2	"	Q	1	1/2	"
E	1	1/2	"	R	1	1/2	"
F	1/2	0	A	S	1	1/2	"
G	1	1/2	A, T, V, W, Y	T	1/2	0	A, J
H	1	1/2	"	U	1	1/2	A, T, V, W, Y
I	1	1/2	"	V	1/2	0	A, J
J	1	1/2	"	W	1/2	0	"
K	1	1/2	A, O, T, V, W, Y	X	1	1/2	A, O, T, V, W, Y
L	1/2	0	T, V, W, Y	Y	1/2	0	A, J
M	1	1/2	A, T, V, W, Y	Z	1	1/2	A, T, V, W, Y

LESSON 4

ORTHOGRAPHIC PROJECTION

(27) **Definition.** *Orthographic Projection,* the branch of *geometry* employed in the making of *working drawings,* may be termed the **"science of proportional drawings."** This definition means little without some further explanation; however, it is perhaps well to give it at this time as a foundation on which to base further discussion.

PRINCIPLES OF ORTHOGRAPHIC PROJECTION

(28) It is seen from Fig. 50, which is a representation of a cube constructed by the principles of *Descriptive Geometry,* applied in what is known as *Perspective,* that, tho the object is represented as we are accustomed to see it, the picture gives us absolutely no conception of the ratio of the several parts of the object to each other; i.e., tho the sides of the small square recess in the top may appear to be half the length of the edge of the cube, one has no means of knowing exactly what the relation is; hence, unless actual dimensions were given for every detail of such a drawing and these dimensions could be depended upon as being absolutely accurate, one would have no means of making, except approximately, the object which the drawing represents. Hence, it will be appreciated that in making drawings for the use of workmen in shops, such an application of *Descriptive Geometry* should be employed as will represent each line of the object at least once, in its *true mathematical ratio* to other lines; i.e., such a representation, that if no

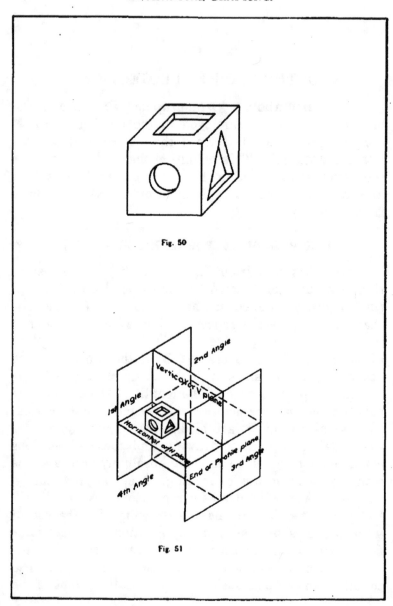

Fig. 50

Fig. 51

dimensions were given, one could compare lines by means of a scale or dividers and be certain of their exact ratio to each other. This branch of *Descriptive Geometry* is known as **Orthographic** or **Proportional Measurement Projection.**

Orthographic projection. To obtain such a projection of the cube represented in Fig. 50, let us imagine that we have suspended the cube in space with the face containing the square recess horizontal; then, see Fig. 51, let us imagine that four *planes* be drawn about this cube in the positions shown, one, a *horizontal* plane, a second a *vertical* plane parallel to the face of the cube containing the circular recess and two other planes *perpendicular* to both the *vertical* and *horizontal* planes just drawn.

Coordinate planes and coordinate angles. The *four planes* just constructed about the cube, Fig. 51, are known in orthographic projection as **coordinate planes** and are named individually, the **Horizontal** or **H plane,** **Vertical** or **V plane, Profile** or **End plane.** The *four diedral angles* formed by the H and V planes are known as **1st, 2nd, 3rd,** and **4th** and are numbered in the order shown.

Projections or orthographic representations. Before proceeding with the explanation of the manner of obtaining the proportional drawings, a fact of geometry should perhaps be called to mind; i.e., the *point* is the *origin* of all *geometric conceptions,* including *lines, planes,* and *solids;* for, if a *point moves thru space* in a *fixed direction* it *generates* a *right line;* if this *line* be *moved thru space* in a *fixed direction* it *generates a plane,* and if this *plane* be *moved thru space* it *generates* what we commonly call a *solid;* hence, in making representations of

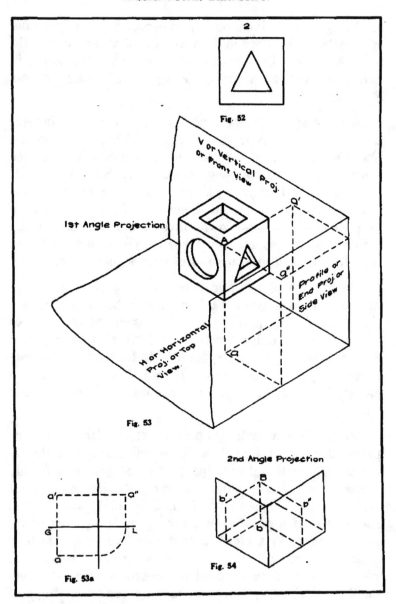

2

Fig. 52

V or Vertical Proj. or Front View

1st Angle Projection

Profile or End Proj. or Side View

H or Horizontal Proj. or Top View

Fig. 53

Fig. 53a

2nd Angle Projection

Fig. 54

any object we find it possible always to simplify the work by making the representations of the various *significant points* of the object and connecting these points by right lines, etc., Fig. 52.

In explaining the method used in obtaining the *orthographic representations* of a cube, the corner A, Fig. 53, will be taken as typical of all significant points of the object. It is desired to represent this point on each of the three coordinate planes, the second End plane being for the time eliminated. From point A are dropped three perpendiculars, one to each of the coordinate planes; the points in which these perpendiculars pierce these coordinate planes are known as the **projections** of point A, and are called, **V** or **Vertical projection** or **Front View** (*always lettered a′* if lettered at all), **H** or **Horizontal projection** or **Top View** (*always lettered a* if lettered at all), and **Profile, End Projection, End** or **Side View** (*always lettered a″* if lettered at all); if then from all of the *points* of the object *perpendiculars* were dropped to the *Vertical plane* and *lines* drawn connecting the *piercing points* of these *perpendiculars* in regular order, Fig. 53, we would have on the *Vertical plane* a *drawing* or *projection* representing perfectly the *appearance* of the *front* of the cube; a similar process would give us on the *Horizontal plane* a correct representation of the *Top* of the object and on the *End plane* a representation of the *side* of the object.

1st, 2nd, 3rd, and 4th angle projections. If the point be placed in the *first angle,* as in Fig. 53, the *projections a, a′, and a″* are known as **First Angle projections.** *Projections b, b′, and b″* of point B in the *second angle,* Fig. 54, are known as **Second Angle projections;** *c, c′, and c″,* the *projections of point C* in the third angle, Fig.

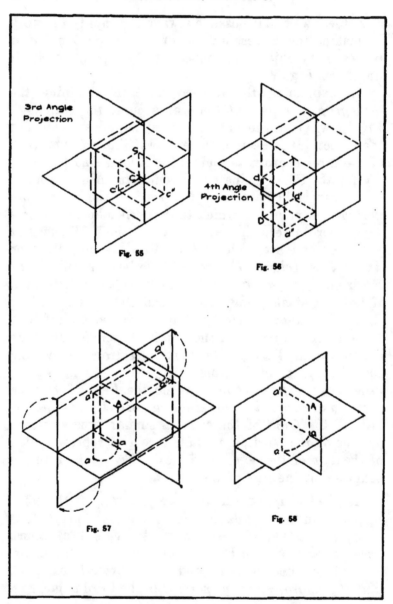

3rd Angle
Projection

Fig. 55

4th Angle
Projection

Fig. 56

Fig. 57

Fig. 58

55, are **Third Angle projections;** and *d, d', and d''* of
point D in the *fourth angle,* Fig 56, are known as **Fourth
Angle projections;** i.e., the *projections* of a *point* are
known as **First, Second, Third,** or **Fourth Angle** projections according to the *angle* in which the *point* is *placed.*

Revolution of coordinate planes. Already an apparent difficulty has arisen in the question of how to represent all of these projections, e.g. of point A, Fig. 53, on a
single sheet of paper when in fact the three projections,
a, a' and *a'',* are found on three planes at right angles to
each other. The line of intersection of the *Horizontal*
and *Vertical planes* is known as the **Ground Line** or **G L;**
the intersection of the *Vertical* and *End planes* is $G_1 L_1$.
This difficulty can now be solved as follows: Using G L
as an axis, Fig. 57, let us imagine that the portion of the
H plane in *front* of *V* is **revolved down,** the portion *behind,* **up** until it *coincides* with the V *plane.* In this revolution, *a,* Fig. 57, revolves into the new position, *a,* on
the continuation of the perpendicular dropped, from *a'*
to G L; for, if thru the two lines A*a* and A*a',* a plane
be passed, Fig. 58, it will cut from the V plane a line
thru *a'* perpendicular to G L; as *a* revolves about G L
down, it revolves in the plane *a'* A*a,* and when it reaches
the V plane, must lie on the perpendicular to G L thru
a', the line cut from the V plane by the plane *a'* A*a.* Since
a is before G L the distance A*a',* *a* will be found below
G L the same distance; *a'* is above G L the distance A *a;*
hence, the distance from G L to the points *a* and *a'* represents the exact distances which the *point,* of which these
are the *projections,* is from the V and H planes. If then,
$G_1 L_1$ be used as an axis and the portion of the *Profile
plane* in *front* of V be revolved to the **right,** *a''* comes
into the new position *a''*, a distance to the right of $G_1 L_1$

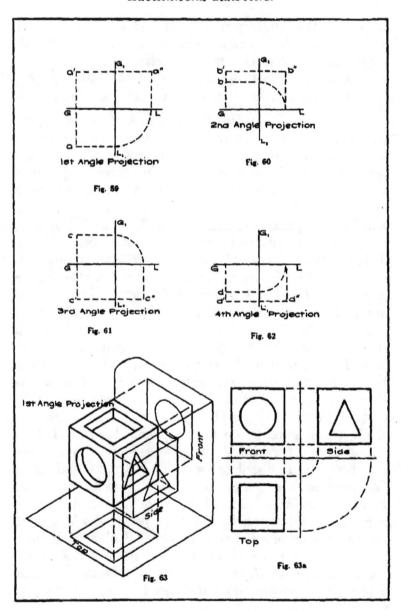

1st Angle Projection

Fig. 59

2nd Angle Projection

Fig. 60

3rd Angle Projection

Fig. 61

4th Angle Projection

Fig. 62

1st Angle Projection

Front

Side

Top

Fig. 63

Front

Side

Top

Fig. 63a

equal to Aa' and a distance above G L equal to Aa, or on the horizontal line thru a'. Transferring these revolved positions to a new figure, Fig. 59, we have the point A in the *first angle* represented by the three projections, a' above G L, a below GL, and a'' to the right of G_1 L_1 and above G L. In a similar revolution the projections of point D, Fig. 56, would revolve into the positions d, d', d'', Fig. 62; however, in this case both d and d' are below G L. By a revolution of the profile plane behind V to the right the projections of point C, Fig. 55, revolve into the positions c, c', c'', Fig. 61; in this case c' is below G L and c above, the reverse of point A in the first angle. For second angle see Figs. 54 and 60.

Projections of objects. Advancing from the points A, B, C, and D just discussed, to the *objects* of which these points are elements, we find the *projections* of the cube, when placed in the *first angle,* to be as in Fig. 63. When placed in the *second angle,* to appear as in Fig. 64, *third angle,* Fig. 65, and *fourth angle,* Fig. 66. When the coordinate planes are revolved in each of these cases the projections revolve into the positions shown in Figs. 63a, 64a, 65a, and 66a.

Elimination of 2nd and 4th angles. If any of these groups of projections or drawings is to be made use of in constructing the object, a glance shows that it is clearly impossible to make use of the two with the object in the 2nd and 4th angles; for, in each case, after the revolution the two projections are one over the other, producing a hopeless muddle. The choice is then necessarily between the *first* and *third* angles.

Elimination of the first angle. From Fig. 67, it is noted that as we ordinarily see objects the *top* appears

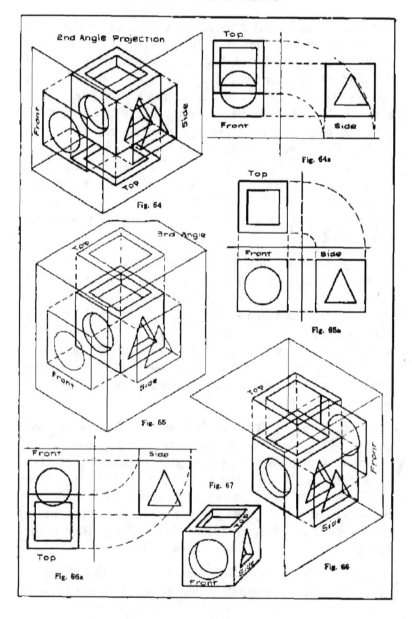

2nd Angle Projection

Front

Side

Top

Fig. 64

Top

Front

Side

Fig. 64a

Top

Front

Side

Fig. 65a

3rd Angle

Top

Front

Side

Fig. 65

Front

Side

Top

Fig. 66a

Fig. 67

Front

Side

Top

Fig. 66

Top

Front

Side

above the *front* and the *right end* to the *right* of the *front* or the *left end* to the *left* of the *front,* according to the position from which the cube is seen. When the object is placed in the *first angle* and the projections revolved into the positions shown in Fig. 63*a,* it is seen that altho the *right end projection* comes in its natural position to the *right* of the *front,* the *top* is **under** the *front,* an arrangement by no means natural. While, when the object is placed in the *third angle,* Fig. 65, and projections revolved as shown in Fig. 65*a,* the views assume a grouping identical with their order on the object itself; i.e., the *right end* to the *right* of the *front,* and the *top* **above** the *front.* Merely for the sake of this natural arrangement the *third angle* will be selected in preference to the *first* in making *working drawings;* i.e., all working drawings will be **third angle orthographic projections.**

(29) **Summary of principles.** It may be well to summarize a number of principles brought out in this discussion, likewise to mention several violations of pure orthographic projection. The *top view,* Fig. 65*a,* represents the exact appearance, with lines in true proportions, of the *top* of the cube; the *front view* represents the same of the *front* of the cube, and the *side view* the same of the *side* of the cube.

The *top* and *front views* **must** be *directly above* and *below* each other and the *front* and *end views* **must** be on the *same horizontal lines* as shown in Fig. 65*a,* if the group is to represent the true orthographic projection of the cube; a violation of this renders the whole drawing incorrect.

(30) **Permissible violations.** In Fig. 57, it is shown that the portion of the *End* or *Profile plane* in

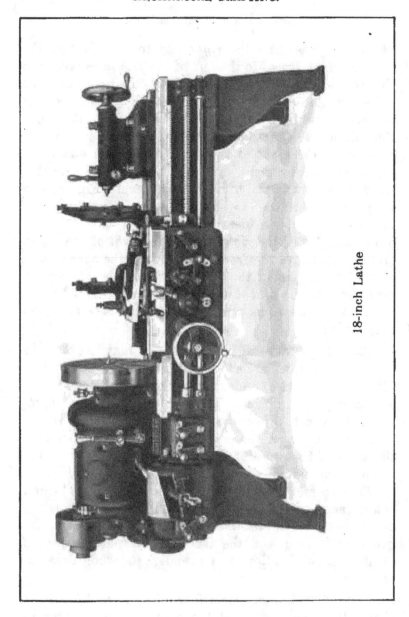

18-inch Lathe

front of V is revolved to the **right**; this of course means that the portion of the *Profile plane behind* V revolves to the **left**; while, in Figs. 65, 65a, the portion of the *Profile plane behind* V is represented as being revolved to the **right**. This revolution to the **right** of the portion of the *Profile plane behind* V is *orthographically incorrect;* however, in the case of the *third angle* projections it is tolerated for the natural order of projections which it produces.

WORKING DRAWINGS

(31) **Definition.** A **working drawing** of a piece of machinery is such a *group of correctly and completely dimensioned orthographic views of that object as will give all the information necessary in construction a duplicate of the same.*

(32) **Detail drawing, defined.** A **detail drawing** is a *working drawing of one piece of any machine.*

Detail signature. Accompanying each *detail drawing,* whether the detail drawing be by itself or one of a set, should be given a characteristic **signature** containing the following information: *Name* of the *machine part, material* of which it is made, *number* of *parts required,* and some *arbitrary number* for the *pattern* if the object is to be cast. This information should be given in the following manner:

Valve Crank—C. I.
Reqd.—1.
Pattern No. A-3.

The letter **A** in the *pattern number* refers to the *sheet* **A** of the Details of the Corliss engine of which this *valve crank* is a part; the number **A-3**, indicates that the *valve crank* is *detail* number **3** on *sheet* **A**.

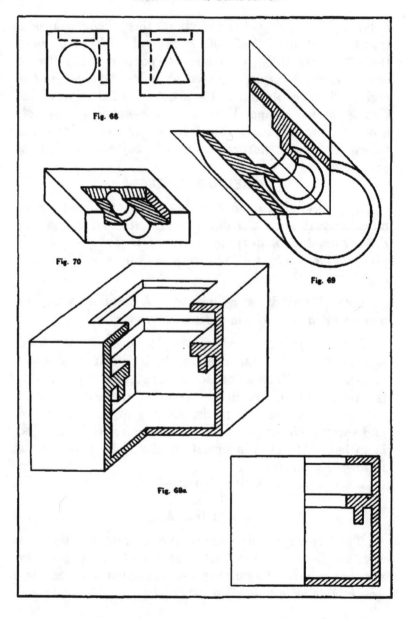

Fig. 68

Fig. 70

Fig. 69

Fig. 69a

SECTIONING

(33) The primary function of orthographic projection is, of course, to represent the portions of an object *visible* to the eye. Any constructions *hidden* by the *surfaces in view* may be represented by conventional lines known as **hidden lines,** Fig. 68. The dotted lines in this figure represent the three recesses in the top, front, and side of the cube. It may be satisfactory to represent in this way the interior construction of so simple an affair as the object shown; however, if the interior construction is in the least elaborate this method is by no means satisfactory. If an interior construction be represented by a number of hidden lines which *cross* each other, the drawing becomes so vague as to be almost unintelligible. For this reason a substitute method has been devised for showing any interior or hidden construction. According to this method it may at any time be imagined that a *cutting plane, parallel* to one of the *coordinate planes,* can be drawn in any position, *cutting away* such portions of the object as *will expose any other parts one may wish to see.* Ordinarily these planes will be found to pass thru some axial line of the object, Fig. 69; however, if desired, they may be imagined drawn elsewhere, Fig. 70. This process of sectioning is purely imaginary and may be represented on only one view of a two or three view working drawing, the other views representing the object unsectioned by any such plane. The process of sectioning is strictly utilitarian; i.e., one should section *only* objects whose construction can be *more clearly explained* by this process than otherwise.

Certain terms applying to sections are sometimes confused from the fact that they may apply either to the drawing or the object. These terms are quarter section, half section, and full section. The confusion ordinarily comes between the terms quarter and half section. From Fig. 69a, it is seen that when one quarter of the object is removed one-half of the drawing will be found sectioned. Likewise if the entire front or half of the object were removed the drawing would be full sectioned. It will thus perhaps be better always to refer to the drawing when speaking of sections; the term quarter section will then, of course, be entirely discarded, as it can refer only to the object.

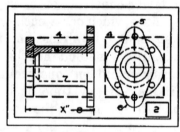

Fig. 71

ORDER OF PENCIL WORK

1—Border Lines
2—Title Space
3—Select Scale
4—View Spaces
5—Center Lines
6—Main Outlines
7—Inside Lines
8—Aux. & Dimen. Lines
9—Sec. Lines & Notes

ORDER OF PENCIL WORK

(34)　The most rapid progress can be gained in the pencil work of a working drawing by following the order given in Fig. 71.

Caution. It is by no means wise to attempt to finish one of several views before doing any work on the others. Fewer mistakes will be made and more rapid progress gained by working on all views at the same time; i.e., when a line is placed on one view, its projections in the other views should be obtained before proceeding with other lines. All views will thus be finished at practically the same time. When one projection of a line is obtained from a given dimension, the other views of this same line should be obtained by the principle of projection rather than by making use of the scale a second time. This practise, tho it may permit a mistake to remain undetected, has the advantage of producing drawings which are true orthographic projections.

LARGE COMPASS

(35)　**Adjustment.** To adjust the *needle point* of a *compass* to both the *pencil* and *pen, remove* the *pencil point* and *insert pen;* after *adjusting* the *needle point* so that its *shoulder,* **not the point,** *is flush* with the *end* of the *pen, remove pen* and inserting *pencil,* adjust *lead* until it is *even* with the *shoulder* of the *needle point.*

Sharpening lead. Sharpen the lead of a large compass the same as the lead of the Bow Pencil. Art. 17, Lesson 2, Fig. 32.

Use. For *adjustment* of *leads* to any desired *radius,* and placing *needle point* at any *desired center,* see **Large**

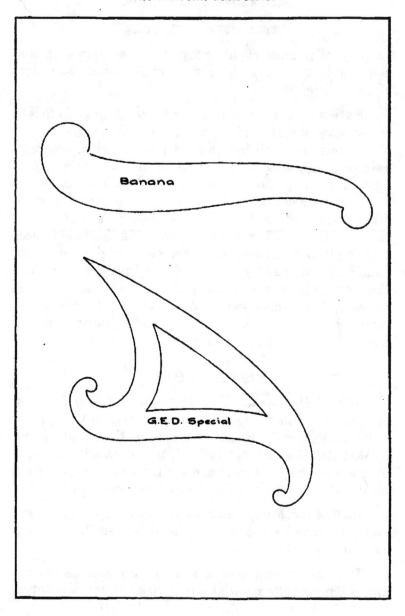

Dividers, Art 14, Lesson 2.   For *describing arcs* see **Bow Pencil**, Art. 17, Lesson 2.

IRREGULAR CURVE

The irregular curves are those which cannot be drawn readily and accurately with the compass.   The general directions of the different portions of such curves are first determined roughly by a number of plotted points at as small intervals as possible (the positions of these points are obtained either by mathematical coordinates or mechanically from other projections or views of the same curve).   Before drawing the curve mechanically it is best to draw lightly a freehand curve thru the plotted points, then carefully piece by piece the mechanical curve may be drawn.   In drawing the mechanical curve two things must be kept in mind; first, that the final curve must coincide as absolutely as possible with the freehand curve; second, that the curve must be "smooth," i. e., it must have no sudden glaring changes of curvature or "Humps."   The failure of a novice to obtain a good irregular curve is due to perhaps two causes: first, that he starts with the assumption that it is too easy to require any attention, and second, that he is too easily satisfied with a very indifferent job.   Curves having curious "humps" may be termed freaks and are seldom, if ever, encountered in Mechanics.

It is difficult to recommend any curve or even several curves as being even approximately universal, so no such advice will be attempted.   A great number of such curves are listed in all instrument catalogs and special requirements will have to be depended upon in any selection. However, two curves have found much favor among students and are recommended for general use.   One of

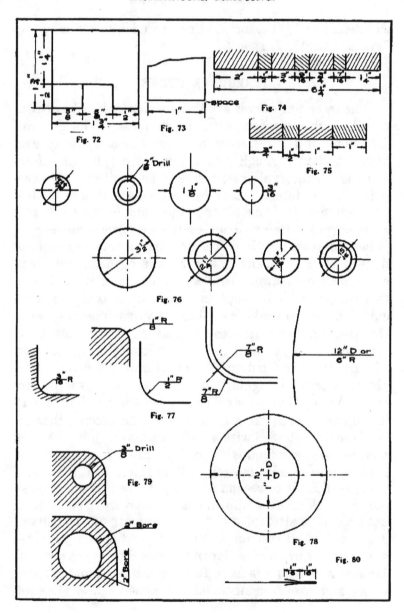

Fig. 72

Fig. 73

Fig. 74

Fig. 75

Fig. 76

Fig. 77

Fig. 78

Fig. 79

Fig. 80

these has obtained the name of "Banana" curve and the other is the G. E. D. Special.

DIMENSIONING

(36) In *dimensioning* the following rules or suggestions should be observed:

(a) **Dimensions** should read from **left** to **right** or **up.** Fig. 72.

(b) The **auxiliary lines** used in dimensioning should *not quite connect* with the lines from which they lead. Fig. 73.

(c) **Series.** A *series* of dimensions should be given on one *continuous dimension line* as in Fig. 74, and **not** as in Fig. 75.

(d) An **overall dimension** should **always** accompany a *series*, both as a *check* and for the *convenience* of the workman. Fig. 74.

(e) **Diameters.** *Diameters* should be placed on a *linear diameter* of the circle or as in Fig. 76 whenever possible; when necessary to indicate the *diameter* on a *straight line projection* of a circle, the dimension should be accompanied by the letter **D.** Fig. 77.

(f) Do not place dimensions *on* or *along Center Lines.* Fig. 78.

(g) Inasmuch as the meaning of *hidden lines* is not always clear, it is bad practice to place dimensions on such lines.

(h) **Leaders.** All *leaders,* Fig. 79, should be made *mechanically* and **not** *freehand.*

(j) **Arrows.** To lessen the difficulties of the beginner in making good arrowheads, the method shown in Fig. 80 is recommended. The arrows are both simple in

Twist Drills

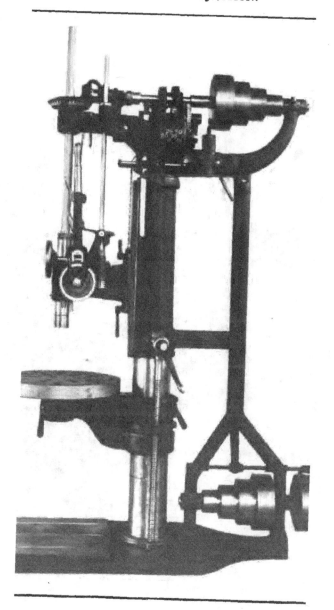

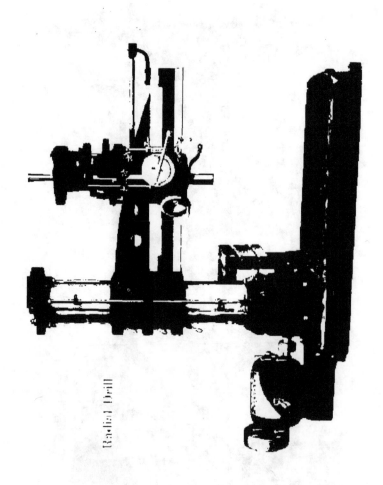

Radial Drill

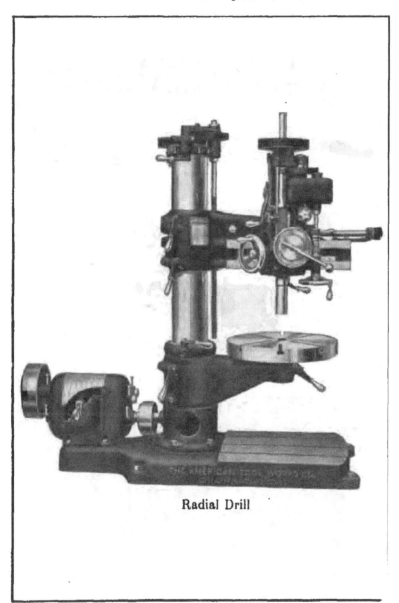

Radial Drill

REAMERS

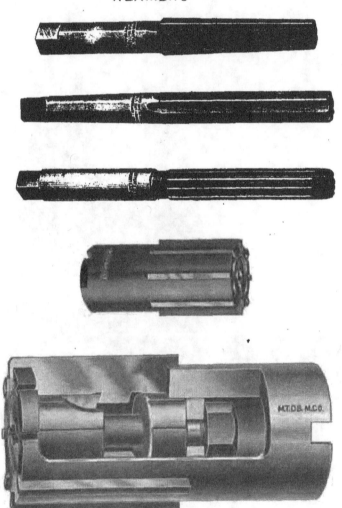

construction and look well. The *arrows* of *dimension lines* contain **two** *barbs,* while those of *leaders,* Fig. 79, but **one.**

(k) **Notes.** For *explanatory notes* the *leader* should end so that the notes may read either *horizontally* or *vertically* as the *dimensions,* but **not** *diagonally.* Fig. 79.

(1) Dimensions up to two feet should be stated in inches; e. g., 12″, 18″, etc.

Two feet may be written either as 24″ or 2′-0″.

Except for sheet metal, dimensions above two feet should be expressed as follows: 2′-3″, 6′-4″, 7′-0″, etc.

The dimensions for sheet metal should be given in inches and in the order of *thickness, width, length;* e. g., ⅛″ x 36 x 120.

SHOP TERMS

(37) **Drill.** Quite frequently instead of indicating the diameter of a hole on the drawing according to the suggestions under *Diameters* in *Dimensioning,* it is found convenient to substitute a *note* which gives at the same time the *diameter* of the hole and the *shop operation* necessary in making that hole. Round holes up to 1¼″ or 1½″ in diameter are ordinarily cut with twist drills such as are shown in Fig. 81. In such cases the note that will be substituted for the diameter is (¾″ Drill, 1″ Drill, etc.). Such drilling operations can be done on a Lathe, tho more conveniently and rapidly on any of the types of Drill Presses shown in Fig. 82.

(38) **Fillet.** It is a well recognized principle of mechanics that a *break* is much more likely to occur in *sharp corners* of a machine than elsewhere, the corner seeming to furnish a starting point for the break. For this reason and others which need not be mentioned, all corners found on castings are seen to be slightly rounded,

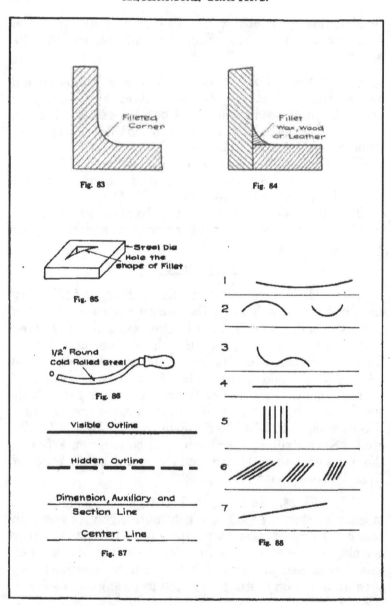

Fig. 83

Filleted Corner

Fig. 84

Fillet
Wax, Wood
or Leather

Steel Die
Hole the
shape of Fillet

Fig. 85

1/2" Round
Cold Rolled Steel

Fig. 86

Visible Outline

Hidden Outline

Dimension, Auxiliary and
Section Line

Center Line

Fig. 87

Fig. 88

Fig. 83. This rounded corner is known as a **fillet**; likewise the *material* which is used to make this *fillet* in patterns takes the same name. In Fig. 84 is shown the method of making such filleted corners in patterns. The triangular piece shown is made of **wood**, shaped by driving thru a Die, Fig. 85, as *dowel pins,* or of **hard wax** rounded by a heated rod, Fig. 86, or it may be of **leather** which can be purchased in coils of any length. The leather fillets, of course, are most convenient for very irregularly shaped pieces. The radius of the arc of such fillets is quite generally $\frac{1}{4}''$; however, it is necessarily a matter of machine design and for very large pieces the radius must be greater than $\frac{1}{4}''$.

(39) **Conventional Lines.** The conventional lines shown in Fig. 87, are standard and should be followed strictly. Concerning the *hidden lines*, it may be said that **no one thing except dimensions will add to or detract from the appearance of a drawing more than care or lack of it in the correct drawing of hidden lines, both as to the uniform length of the dashes and uniform space between dashes.** In *tracing,* follow strictly the *weights* given in the figure for these various conventional lines.

(40) **Order of inking in tracing.** In *tracing,* the following *order* should be observed for most rapid and accurate work, Fig. 88.

1. Large circles and arcs.
2. Small circles and arcs with the bow pen.
3. Irregular curves with special curve.
4. Horizontal lines with the T-square.
5. Vertical lines with T-square and triangles.
6. Inclined lines in groups, e.g., 30°, 45°, and 60°.
7. Other oblique lines.
8. Dimension and auxiliary lines.
9. Section lining, dimensions and notes.

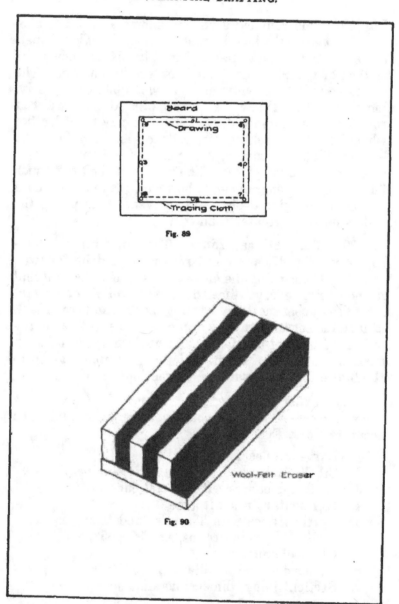

Fig. 89

Fig. 90

TRACING CLOTH

(41) **Tracing cloth** is a medium quality of **linen** coated with a preparation which gives it a smooth hard surface and renders it transparent. Of the various grades of cloth on the market, the imported brand "**Imperial**" gives by far the greatest satisfaction and is recommended for general use. Always before making use of any piece of cloth be sure to rip off about ⅛" of the *selvage* edge; it may prevent a bad buckling of the tracing.

Tacking to the board. It is best always to have the sheet of tracing cloth *slightly larger* than the sheet of paper so that the tacks used in pinning down the cloth may be placed *outside* the sheet. In tacking down the cloth, preferably always with the *dull side up,* be sure to stretch very tight and tack firmly. See Fig. 89 for best order of tacking.

Preparation of the surface of the cloth. Unless prepared in some way, the *surface* of the tracing cloth will likely take the ink very poorly, giving ragged and faded out lines. The cloth may be dusted or rubbed with **chalk** or preferably (**magnesium carbonate**), which may be purchased at any drug store in 5-cent blocks, and then rubbed with a piece of linen. A draftsman may find it much to his advantage to have in his kit for the dusting of the surface of the cloth, a new (wool felt) blackboard eraser, Fig. 90. If kept full of chalk or magnesium dust, the eraser gives most excellent results.

Order of work. Unless one is sure of being able to finish the tracing of several views of a drawing before it is necessary to stop work, it will be found best always to *trace one view at a time, finishing* that view *before leav-*

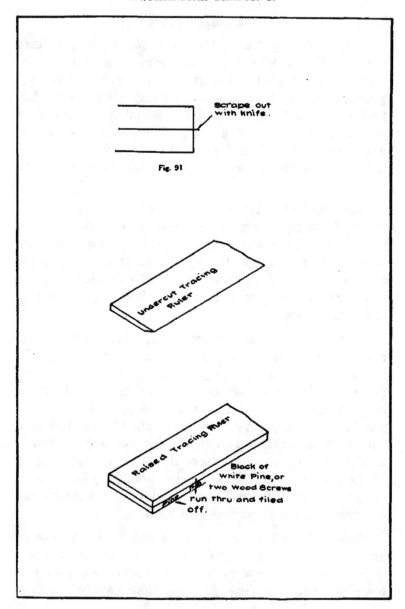

Scrape out with knife.

Fig. 91

Undercut Tracing Ruler

Raised Tracing Ruler

Block of White Pine, or two Wood Screws run thru and filed off.

ing work. Tracing cloth has a decided tendency to stretch and warp, and it may be found most difficult to make old lines check with new if the tracing has been left standing for a day or more.

Erasing. In erasing, use the *pencil eraser* always in preference to the *ink eraser* or *knife.* It may require more time to erase a mistake; however, the cloth will be found in good condition after such erasing; while the *ink eraser* or *knife* quite easily roughens the surface and causes blots on application of new ink. A *knife* may be used to advantage in *scraping out* slight *accidental extensions* of *lines,* Fig. 91.

Caution. If necessary to rule *across* ink lines, be sure to move the pen **rapidly.** If the pen is moving slowly the ink will likely follow down the old ink line to the T-square or triangle and cause a bad blot.

Weights of lines. It will be necessary to use but *two* weights of lines thruout the work in tracing. A **number 3** line, slightly less than 1/32″ for all *outlines,* both *hidden* and *visible,* and a **number 1/2** line (a very thin fine line) for all *dimension, auxiliary, section,* and *center* lines.

BEVELLED AND RAISED TRACING RULES

For tracing the two rulers shown in Figs. 91a and 91b are indispensable. With the *bevelled ruler,* Fig. 91a, it is possible to rule across inked lines without any danger of blotting. The *raised ruler* of Fig. 91b saves an immense amount of time, as it makes it possible to continue tracing no matter how many ink lines are still wet.

SCALE

(42) In using the ordinary *architect's scale,* which has been designed to make drawings of such size that

½″, ¾″, 1″, etc., on the drawing is equal to 1′ on the object, the beginner may experience some difficulty if he is attempting to make a drawing to the scale of ½″, ¾″ or 1″, etc., to the inch. On inspection of the scale it is found that for each scale the ½″, ¾″, etc., at the end is subdivided into four parts, and each of these is further subdivided into three, six, or twelve parts. One of the four parts represents one-fourth of an inch.

The following table may be of service to the beginner in obtaining from the various scales the dimensions most frequently used.

Scale of 1½″ to 1″ from the 1½ scale.

1/4″ equals space from 0 to 3.
1/8″   "   1/2 space from 0 to 3.
1/16″ "   3 of smallest divisions.
1/32″ "   1½ smallest division.

Scale of 1″ to 1″ from 1″ scale.

1/4″ equals space from 0 to 3.
1/8″   "   6 of smallest divisions.
1/16″ "   3 of smallest divisions.
1/32″ "   1½ smallest division.

Scale 3/4″ to 1″ from 3/4 scale.

1/4″ equals space from 0 to 3.
1/8″   "   3 of smallest divisions.
1/16″ "   1½ smallest division.
1/32″ "   1½ smallest division on 3/16 scale.

Scale of 1/2″ to 1″ from 1/2 scale.

1/4″ equals 1 of four largest divisions.
1/8″   "   3 of smallest divisions.
1/16″ "   1½ smallest division.

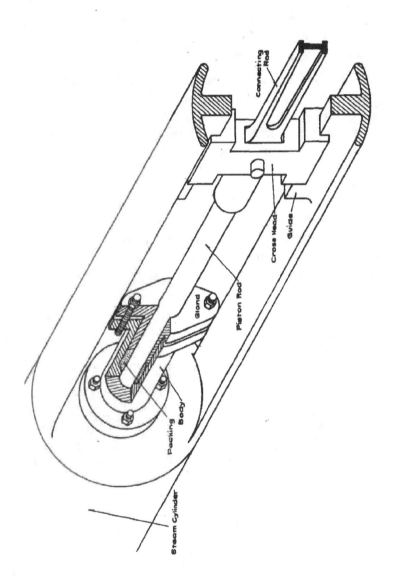

A mechanical engineer's scale has recently been put on the market. Mechanical draftsmen will find it much to their advantage to own one of these.

(43) **Use and construction of the stuffing box.** One of the few uses of the type of *stuffing box* used in the drawing plate of this week is shown in Fig. 94. The stuffing box complete is composed of six parts: **Body, gland,** two **stud bolts,** two **nuts.** As seen, the **body** is bolted to the end of the *engine cylinder,* the *piston rod* passing thru both the *body* and the *gland.* Around the *piston* and between the *gland* and the *shoulder* of the *body* is shown the **packing** (hemp or especially prepared packing) *which,* when the *gland* is drawn down tight by the two *nuts,* is jammed tight around the *piston,* preventing the escape of steam from that end of the cylinder.

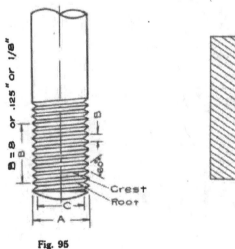

Fig. 95

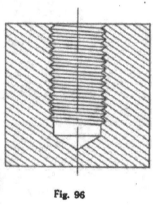

Fig. 96

LESSON 5

U. S. S. THREADS

(44) **True representation.** To represent threads exactly as they appear on a threaded rod or bolt or in a threaded hole would be a very tedious process, for the sharp edge of the thread constitutes what is known in mathematics as the *spiral* and in this case of so slight curvature as to make exact construction most tedious. Hence, in the original attempt to simplify the construction in drawing such threads, *straight lines* were substituted for these *curved spiral lines* and threads were represented as shown in Fig. 95, for threads on a bolt, and as in Fig. 96, for threads in a threaded hole. In each of these cases the notched edges represent the threads as they actually appear in profile.

Crest, root, outside diameter, root diameter, pitch. The *sharp edge* of the thread is known as the **crest**, Fig. 95, the *depression line* as the **root**. The *diameter* of the crest of the thread is known as the **outside diameter** or the **diameter** of the **bolt**. The *diameter* of the *root* is known as the **root diameter** or **diameter** of the **tapping drill**.

For bolts and screws of various sizes the size of the thread, hence, the number of threads per linear inch must vary. For each sized screw there is a standard thread which is indicated by the term **pitch**. This term *pitch* may mean either the *number* of threads *per linear inch*, Fig. 95 (B in this case equals 8, including 7 full threads and two half threads) or, the *distance* in *inches*, either fractional or decimal, *between two consecutive*

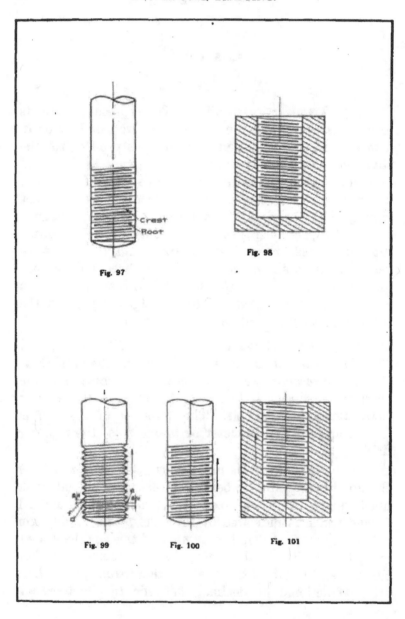

Fig. 97

Fig. 98

Fig. 99

Fig. 100

Fig. 101

thread crests or the *width* of *one thread,* as indicated by
B, Fig. 95.

Conventional representation. Inasmuch as in making
a drawing there is ordinarily no necessity for the draftsman to go to the extra trouble to represent bolt threads
as shown in Fig. 95, the conventional representation,
shown in Figs. 96, 97 and 98, has been devised and is
universally used. In this conventional representation it
is seen that the *light inclined lines* represent the *crest* of
the thread while the *short heavy lines* between represent
the *roots* of the threads, the notches at the right and left
having been omitted.

Slope of thread. In passing around a right-hand
threaded bolt, Fig. 99, moving in a *clockwise* direction
from point *d* to *e,* the moving point has advanced along
the axis of the bolt in a direction shown by the arrow
and has moved in this direction *one-half* the *width* of *one
thread,* or ½ B. Hence, in representing threads on the
front of a threaded bolt, the direction of the slope of the
lines representing the crests and roots of the thread must
be from **left** to **right up** in the direction of the *axis* of
the bolt, Fig. 100.

In moving from point *e* in a *clockwise* direction, to
the point *f,* the moving point has advanced along the
axis in the direction shown by the arrow, a distance equal
to *one-half* the *width* of *one thread* or ½ B. Hence, in
representing the thread in the *back* of a threaded hole,
the lines representing the crests and roots must slope
from **right** to **left up** in the direction of the *axis,* Fig. 101.

Shop note. The following note may be used to indicate the *outside diameter* of the thread and the *pitch* or
number of *threads per inch* for any threaded hole on a

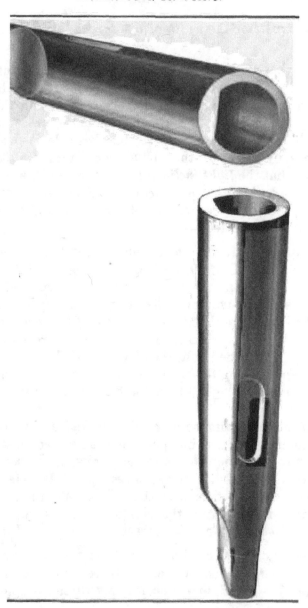

piece of machinery (¾″x10 *pi.*), *pi.* being the abbreviation for *pitch*.

(45) **Tapping drill.** The dimension C, Fig. 95, gives at once both the *root diameter* of the threads, and the *diameter* of the *tapping drill* or twist drill that would be used in drilling a hole to be threaded to accommodate the screws. It is seen that this dimension C, the diameter of the tapping drill, gives the distance between the two parallel lines which limit the notches of the thread.

SCALES

(46) **Architect's scale.** The inches on the *architect's scale* are divided into *halves, quarters,* etc., i. e., into divisions which are multiples of *two,* making it possible to draw, without any interpolation, plans, etc., of objects whose dimensions are given in *feet* and *inches*.

Engineer's scale. On the *engineer's scale* the *inches* are divided into various numbers of subdivisions, these numbers being multiples of ten; i. e., the *inches* are divided into 10, 20, 30, 40, 50, or 60 divisions. By use of this scale without any interpolation maps may be made and drawings plotted directly from field notes in which the distances are all given in *feet* and *tenths* of *feet*.

Scale versus size. A drawing made to such a size that **one-half inch** on the *drawing* equals **one foot** on the *object* drawn, is said to be made to **one-half scale.** However, if the drawing be made so that **one-half inch** on the *drawing* represents **one inch** on the *object* the drawing is said to be made **one-half size** or to a **scale** of ½″ to 1″.

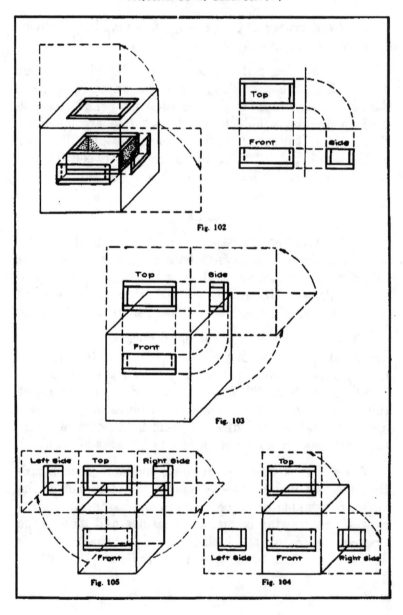

Fig. 102

Fig. 103

Fig. 105

Fig. 104

POSITIONS OF THE THIRD AND FOURTH VIEWS OF A WORKING DRAWING

(47) In Fig. 102 is shown the correct arrangement, orthographicaly, of the three views of a working drawing. However, occasions may arise in which it will be inconvenient to place the side view directly *opposite* the *front;* in this case we may imagine that the line of intersection of the end plane and the horizontal plane becomes an axis about which the *end plane* is revolved, Fig. 103, until it coincides with the *horizontal plane.* This entire horizontal plane is then revolved about its line of intersection with V as an axis, until it coincides with V. The side view will now be found *opposite* the *top* instead of the *front.* If two side views are necessary to show the construction they may be placed on *either side* of the *front view,* Fig. 104, or on *either side* of the *top view,* Fig. 105. No other arrangement is permissible.

In constructing a three view working drawing it is best always to construct the *top* and *front* views from *dimensions* and by *projection;* then, to obtain the *side* views from these two, entirely by *construction* and **not** by the use of *dimensions.* For the sake of construction the two *ground lines,* Fig. 102, may be drawn in lightly; however, they should be erased when no longer needed.

SHOP TERMS

(48) **Tapping drill.** A *tapping drill* is a *twist drill* of the common type, named a **tapping drill** in this case because it has been used in drilling a hole which is to be threaded to receive a screw.

Tap. A **tap** is an instrument, somewhat resembling a bolt, that is used in *cutting threads* in any *drilled hole.*

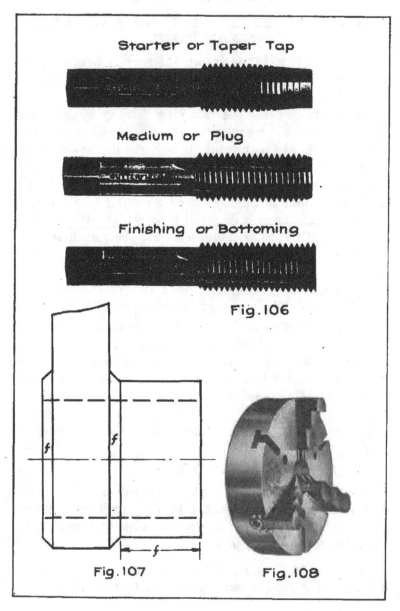

Starter or Taper Tap

Medium or Plug

Finishing or Bottoming

Fig.106

Fig.107

Fig.108

A tap is made by first threading a rod of tool steel as tho it were to be made into a bolt, then grooves are cut or milled lengthwise thru these threads, Fig. 106; on the ordinary type of tap four such grooves are milled, producing four cutting edges. These grooves likewise furnish space in which the chips or shavings may collect. Taps ordinarily come in sets of three, Fig. 106, the one known as a **starter** being ground down on the end to permit it to start easily in the hole. The **medium tap**, ground down only slightly on the end, finishes the thread nearly to the bottom of the hole, while the **finishing** or **bottom tap** is used to finish the thread entirely to the end of the hole.

Finish. To indicate that any *surface* of a piece of *machinery* is to be *finished* or *machined* on a *lathe, shaper,* or *planer,* the letter *"f"* is used, Fig. 107. If this letter *"f"* is omitted on a drawing the workman will understand that a certain surface or surfaces are to be left *rough,* if the piece is a casting or forging.

FACE PLATE

(49) **Chucks versus face plate.** In Fig. 123, which is of a common *lathe,* is shown a piece of material held in place by the four jaws of what is known as a **chuck.** It is seen that these jaws are placed in pairs and are run in and out by means of a screw turned by a key. This *chuck* can be used to hold all pieces of material of a regular shape and convenient size; however, occasion frequently arises in which it cannot be used. In such cases a plate known as a **face plate,** Fig. 109, is screwed on the end of the shaft of the lathe in place of the chuck. The piece of material is then suspended between the two centers indicated, and caused to revolve with the face plate

Fig.109 Motor Driven Lathe

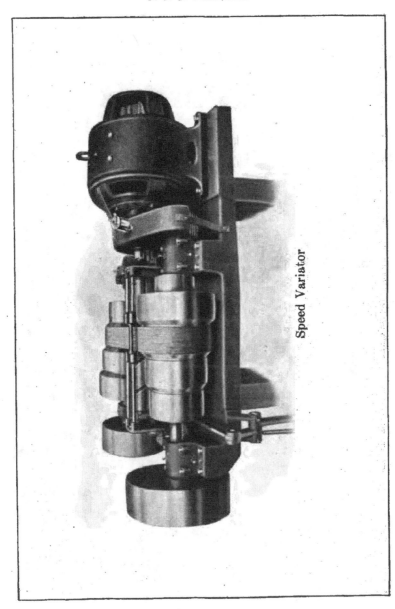

Speed Variator

Fig. 113

by means of various kinds of clamps, Fig. 110. The piece of material may likewise be clamped to the face of the plate by bolts and plates, the heads of the bolts being slipped down into the T grooves seen in the rim, Fig. 111.

(50) **Boring bar.** For outside turning on a lathe the ordinary type of cutting tool is used, Fig. 112. However, for inside cutting or boring a tool known as a **boring tool** is used and if the piece to be bored is a cylinder of any length the tool is held by a bar known as **a boring bar,** Fig. 113.

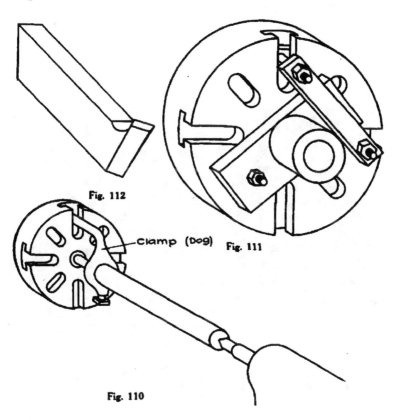

Fig. 112

Clamp (Dog)   Fig. 111

Fig. 110

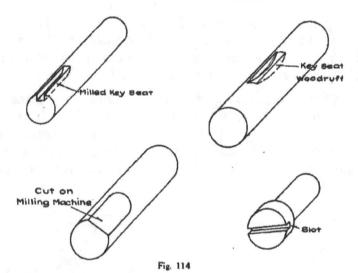

Fig. 114

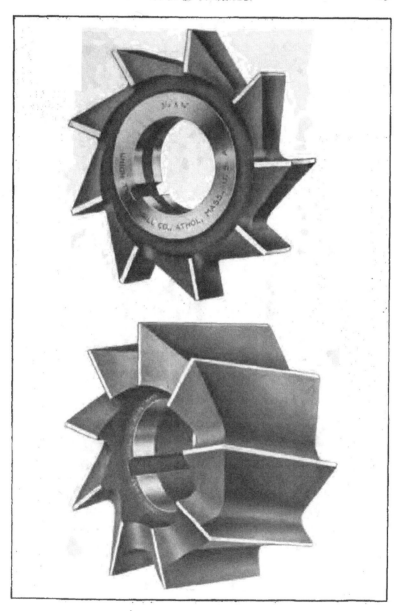

LESSON 6

SHOP TERMS

(51) **Mill.** In the shop operation required in cutting *slots, grooves* known as **key seats,** also other similar operations, Fig. 114, a machine known as a **milling machine** is used. The cutting tools resemble the common type of circular saw and operate on the same principle. As seen in Fig. 115, which is of a **horizontal milling machine,** the cutter is fastened rigidly to the revolving shaft or **arbor** while the piece of material to be machined is clamped to the table and the table moved either by hand or automatically slowly under the cutter, as a log is fed into the saw of a sawmill. For special work milling cutters of many odd designs are made, as seen in Fig. 116. The machine shown in Fig. 117 is known as a **vertical milling machine,** the shaft or arbor in this case being **vertical.** To prevent vibration in the arbor of the horizontal machine (this vibration being known as **chatter**) and consequent rough work of the cutter, **chatter braces,** shown in Fig. 115, are being put on most machines of late design. The note used to indicate any desired milling operation may be as follows: "2″ mill"; the two inches indicating the *diameter* of the *milling cutter;* or *"mill* ⅜″ *key seat, 4″ long," "Mill* ⅛″ *slot,* ¼″ *deep,"* etc.

Tap. In cutting standard threads in nuts or holes which are to receive machine or cap screws, the threading tool, known as a *tap,* Fig. 106, is used. The note that will be used in this connection is ⅝″ *tap,* ¾″ *tap,* etc., or ⅝″x12 *pi. tap.* The ⅝″ or ¾″ dimension in either case gives the *outside diameter* of the thread and

Planer

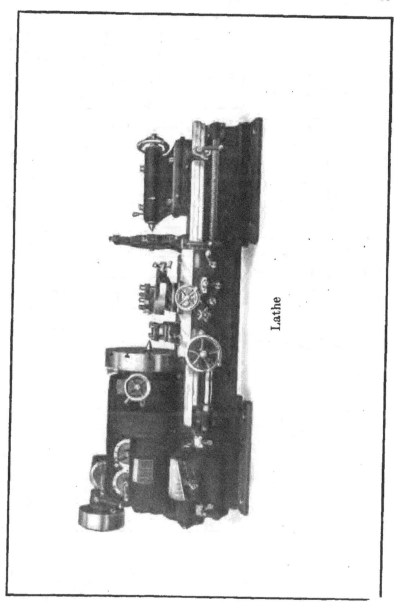

Lathe

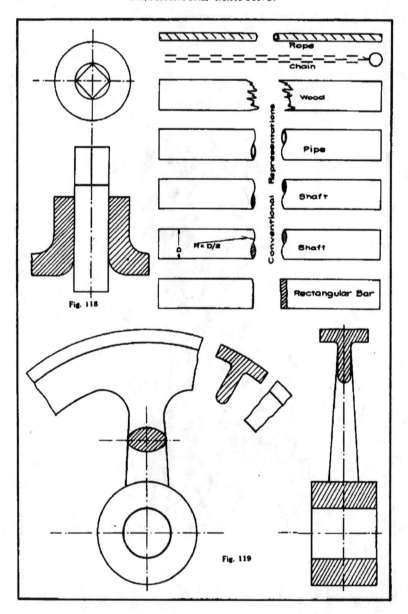

Rope

Chain

Wood

Pipe

Shaft

Shaft

Rectangular Bar

Conventional Representations

$R = D/2$

Fig. 118

Fig. 119

the *pitch* is to be understood as *U. S. Standard;* if not standard, the pitch is to be indicated by the 12 *pi.*, etc., as in the second example.

Bore. In all cases where a round hole is to be machined and the hole is either so large that a twist drill cannot be used or it is desired to give such a finish to the hole as is impossible in the inevitably somewhat rough work of the twist drill, the work will be done on a lathe by means of the short boring tools or by cutting tools in connection with the boring bar and the operation will be termed **boring** instead of drilling. The note referring to such an operation is 7″ *bore*, etc., the *dimension* referring to the *diameter* of the hole. Such boring operations are ordinarily necessary on holes whose diameters are greater than two inches. Twist drills larger than two inches in diameter are not in very common use as it requires an extremely heavy drill press to operate them satisfactorily.

SECTIONING

(52) **Solid cylinders.** The draftsman should keep in mind the fact that there is but one thing to be gained in sectioning, i. e., to show more clearly the *interior construction* of any piece of machinery; if the section does not accomplish this purpose it is just so much wasted labor. This point refers particularly to *solid cylinders*, e. g., *shafts, bolts, screws,* etc., Fig. 118, which should never be sectioned.

Interpolated or revolved sections. In such cases as are shown in Fig. 119, with respect to *rims* and *spokes* of *wheels, standard construction iron,* etc., sections known as **interpolated or revolved sections** are given to show the cross-section of the material at certain places.

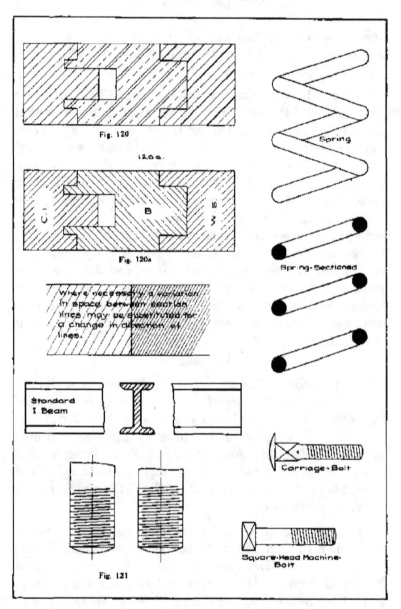

Fig. 120

120a.

Fig. 120a

Where necessary a variation in space between section lines may be substituted for a change in direction of lines.

Spring

Spring-Sectioned

Standard I Beam

Carriage-Bolt

Square-Head Machine-Bolt

Fig. 121

It is a rule that even when a second view of a wheel is given the *spoke* is **not** to be sectioned while the *rim* and *hub* may be if desired, and the shape of the spoke given by two interpolated sections as shown.

Section lines. *Indication of material by variation in section lines.* It is the custom in many shops to indicate the material of which a piece of machinery is to be made by using a characteristic section line for any parts of that piece which have been sectioned, Fig. 120. There are some apparent disadvantages in this, however, for there is at present no universal standard system of section lining. Some shops use one characteristic for *brass, wrought iron*, etc., and others a radically different characteristic. Furthermore, unless one uses these section lines constantly or has a chart of them with him always, he may find it quite difficult to remember some of them. A third objection is that it requires an excessive amount of time to draw some of these section lines.

Indication of materials by abbreviations and universal section lines. For greatest convenience and ease, both in making and reading a drawing, the writers approve the universal section lining with *material* abbreviations as a substitute for the above system, i. e., the use of the standard section line now used for cast iron as the standard for all materials and the particular material of which the piece is to be made indicated by its characteristic abbreviation as shown in Fig. 120a. These abbreviations are easy to remember and the section lines can be drawn rapidly.

THREADS

(53) **Standard conventions.** Both conventional methods of indicating U. S. S. threads, as shown in Fig. 121, are standard and may be used as preferred.

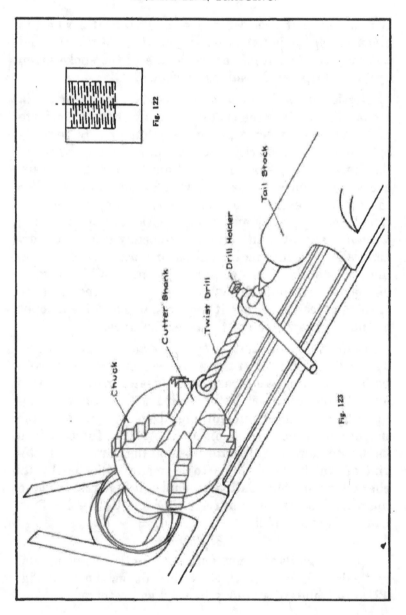

Fig. 122

Fig. 123

Tail Stock

Drill Holder

Twist Drill

Cutter Shank

Chuck

Horizontal Return Fire-Tube Boiler

Fig. 124

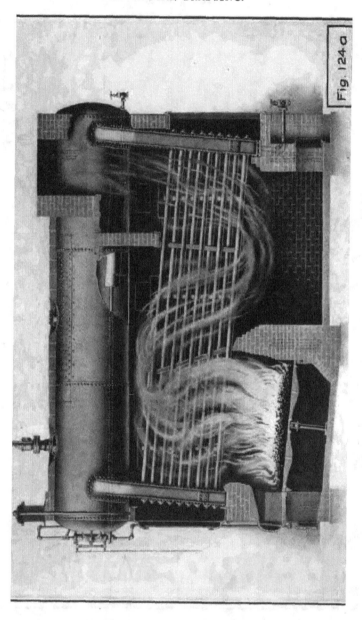

Fig. 124 a

Hidden threads. In representing *hidden threads* in a threaded hold, Fig. 122, the *slope* of the broken lines is the same as that of lines representing threads on the *front* of a threaded bolt.

FLUE HOLE CUTTER

(54) **Process of construction.** The *flue hole cutter* used in the drawing plate of this week is composed of two main parts, **shank** and **tool holder**. The *shank* is made from a bar of machine or *cold rolled steel,* the long taper being turned down on a lathe and the threaded end turned down and the threads *chased* on a lathe. The two flats on the upper end are cut on a milling machine and the hole in the lower end drilled in a lathe, the shank being held in the chuck and the drill in place by the center of the tail stock, Fig. 123. The *tool holder* is forged from a block of *tool steel* (tool steel being used because of excessive strain which the holder must stand). The large round and two small holes were drilled on a drill press and threads cut with a tap. The irregular hole was first forged out roughly and then shaped up on a shaper or filed up by hand.

Use. There are in general use two types of boilers. In the one known as **fire tube** the water is on the *outside* of the *tubes,* and *heated gases* and *fire* pass *thru* the *tubes*. In the other, which is known as the **water tube boiler,** the water is on the *inside* of the *tube* and the fire and heated gases on the outside, Fig. 124. The flue hole cutter mentioned above is used in cutting the holes in the heads or tube sheets for the *tubes* of the *fire tube* type of boiler. Holes ¾″ in diameter are first drilled in the tube sheets for the center or **guide pin** of the cutter Fig. 125; the tool having only one cutter requires this center pin to hold it rigidly in place. See Fig. 125 for operation of cutter.

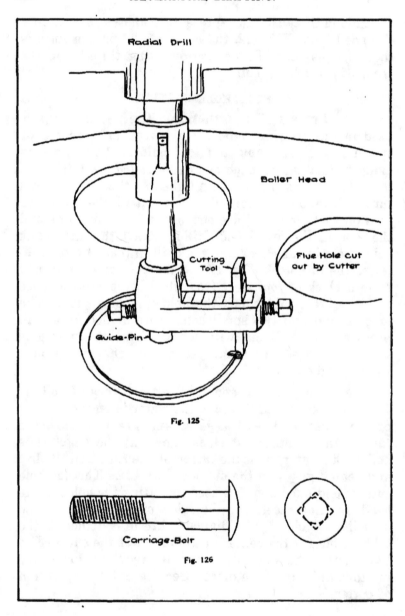

Radial Drill

Boiler Head

Cutting Tool

Flue Hole Cut out by Cutter

Guide-Pin

Fig. 125

Carriage-Bolt

Fig. 126

LESSON 7

STANDARD BOLTS AND NUTS

(55) **Bolts and nuts.** A bolt may be broadly defined as a *round rod of iron or steel, having a head on one end and threaded on the other to receive a nut.*

There are in use at present two distinct classes of bolts, named from their distinctive uses, the one class called **machine bolts** and the other **carriage bolts.**

It is rather difficult to define the term *machine bolt* because of the many shapes such bolts may have and the number of uses made of them. However, a **carriage bolt** *may* be easily defined as a bolt which has an *oval head and whose shank is squared for a distance of from* $\frac{3}{8}''$ *to* $\frac{3}{4}''$ *just under the head;* the side of this square being about the same as the diameter of the remainder of the bolt. This bolt is used in wood work and when drawn into a hole the square under the head takes a grip on the wood which prevents *turning* of the bolt when the nut is drawn. The head being oval and very thin draws down well, leaving very little metal projecting. The bolt will be understood from inspection of Fig. 126.

(56) **Process of manufacture.** The process of manufacture of both general classes of bolts is the same. Rods of iron or steel are cut into pieces of some definite lengths, according to length of bolt desired. These pieces, heated at one end in a furnace, are placed one by one between the two jaws of a machine called a **bolt header,** leaving a certain length of the heated end projecting, and the **ram** of the machine is brought against this heated

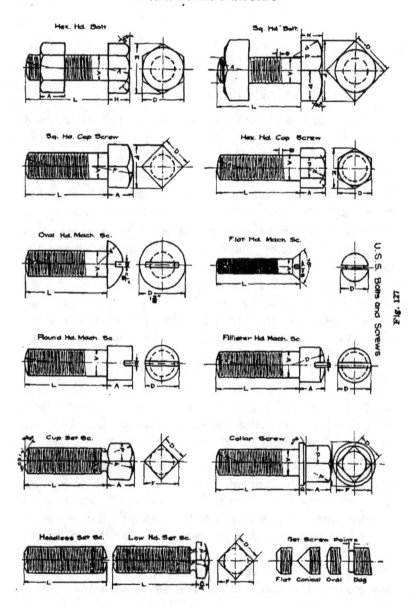

Fig. 121   U.S.S. Bolts and Screws

end with sufficient force to mash and form the heated metal into the desired shaped head. The other end is afterwards threaded in a threading machine. The threading tool, called a ·die, is made of *tool steel,* and resembles somewhat an ordinary nut used on these same bolts.

The nuts for the bolts are either punched from sheets of metal of the proper thickness or cut from steel bars of the proper shape. The ordinary cheap grade of nuts is punched from a sheet of metal, a round hole punched thru the center and this hole afterwards threaded with a tap. Better grades of nuts, usually hexagonal, are cut from hexagonal bars. Holes are punched or drilled in these pieces and these holes threaded as explained above.

(57) **Machine bolts.** Machine bolts are divided into a number of classes, each class being named either from its peculiar *shape* or its distinctive *use.* The dimensions of the several parts of all such bolts have been standardized, tables arranged, and the construction and size of every part of any particular size bolt is perfectly definite.

(58) **Hexagonal and square-headed bolts.** Inasmuch as these two classes of bolts are usually dealt with in the same table of dimensions, it will perhaps be as well to include both in this discussion. In Fig. 127 is shown the conventional manner of representing hexagonal and square-head bolts in a mechanical drawing. It will be noticed that the *hexagonal* bolt is so placed that *three* faces of both the bolt *head* and the *nut* are visible and the *square* head bolt is so placed that *two* of its faces are visible. This placing should be strictly adhered to, especially in machine sketching where it may be necessary to show the kind of bolt by only one view.

Likewise it will be seen that on both bolts the outer corners of the heads and nuts have been ground or turned off until the face of the head and nut is a circle tangent to the hexagonal or square limits of the head or nut. This bevel on the head or nut is called the **chamfer** of the head, etc.

In constructing the *end view* of any bolt the *chamfer circle* is first drawn (the diameter of this chamfer circle will be found in table under head of "*Distances Across Flats*" or "*Short Diameter*") and the *hexagon* or *square* circumscribed by means of the 30°x60° or 45° triangles. **No other method should be used for obtaining the hexagon or square.**

The **length** of such bolts is always the *distance* from the *end* of the *bolt* to the *under surface* of the *head*.

For each different diameter of bolt a standard thread has been selected and named according to number of threads per linear inch; e. g., on a standard ⅝″ bolt will be found 11 threads per inch. The *number of threads per inch* is usually spoken of as *pitch*; e. g., *11 pitch* means *11 threads per inch;* the word *pitch* is usually abbreviated to *pi.,* and a bolt may be explained by some such note as ¾″x10 *pi.,* meaning a bolt ¾″ in diameter, having 10 threads per inch.

In listing such bolts in a *Bill of Materials,* the following order should be used: *1x8x4, Fin. Hex. Bolt.* This indicates a finished hexagonal 1″ bolt, 4″ long, 8 threads per inch or 8 pitch.

The geometrical construction indicated in Fig. 127 is the conventional construction and should be followed carefully.

(59) **Cap-screws.** Cap-screws, both hexagonal and square, are intended for *fasteners;* e. g., to hold a cylinder head to the cylinder, etc., in which case the screw passes thru a hole in the cylinder head and screws into a threaded hole in the cylinder. However, the cylinder is mentioned merely by way of illustration, stud bolts being mostly used in this particular way, acting partially as guides in assembling. Fig. 127.

All cap-screws of 1″ and less in diameter and 4″ long and under, are threaded *three-fourths* of their length; when longer than 4″, they are threaded *one-half* length. They should be listed in Bill of Materials as follows: ½″ — 12x1½″*Hex. Hd. Cap-Screw.*

The geometrical construction for these screws will be found on sheet of Bolts and Nuts, Fig 127.

(60) **Set-screws.** Set-screws are named from their *use* and further divided into different kinds of set screws according to the *shape* of the *point* or *head.* All set-screws may be classed as *fasteners,* being used to clamp or hold one piece of machinery in a definite position with respect to another; e. g., to hold a pulley rigidly to a shaft, etc. Such screws are either made of *tool steel,* oil hardened, or of *machine steel* or *wrought iron,* case hardened. No screw whose *head* exceeds the *diameter* of the *body* more than 1/16″ should be classed as a *set-screw.*

From the figure on the Bolt Sheet, Fig. 127, it will be seen that the *length* of the *headless set-screw* or *gib-screw* is the *total length* of the screw.

Set-screws should be listed in a Bill of Materials as follows: ½″ — 12x2″ *Set-Screw.*

(61) **Collar-screws.** In many cases to prevent scarring from friction of the head of a cap-screw, washers are needed. A screw known as the **collar-screw** combines

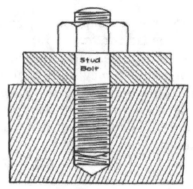

Fig. 128

Shaper with Universal Table

this washer with the square head. This variety of screw will be found mostly on machines whose parts fit snugly and require only a moderate clamp, or where the screw is frequently being loosed and drawn. Example will be found on Bolt Sheet, Fig. 127.

This screw should be listed in Bill of Materials as follows: ½″ — 12x2″ *Sq. Hd. Collar-Screw.*

(62) **Round and fillister head machine-screws.** Quite frequently it is necessary to use screws in places in which it is difficult to use a wrench. For such uses screws with slotted heads are made and named **round** or **fillister head machine-screws** according to the construction of the *head;* examples may be found on Bolt Sheet, Fig. 127.

These screws are made from bars slightly larger than the heads, cut to proper lengths, turned down and threaded in screw machine, and should be listed in Bill of Materials as follows: ½″ — 12x1½″ *Rd. Hd. Machine-Screw.*

(63) **Flat-head machine-screws.** Flat-head machine-screws have *countersunk heads* and are to be used where it is desired to have the heads flush with the surface of the piece into which they are screwed. The holes into which such screws are drawn must be countersunk to receive the heads. See example on Bolt Sheet, Fig. 127.

Such screws should be listed in a Bill of Materials as follows: ¾″ — 10x2″ *Flat-Hd. Machine-Screw.*

(64) **Stud Bolts.** A *stud bolt* is a rod of iron or steel threaded on both ends. It is used largely as in the case of cylinder heads where it is desired that the bolts shall act as guides for placing the head quickly and easily into position; the stud is screwed tightly into the

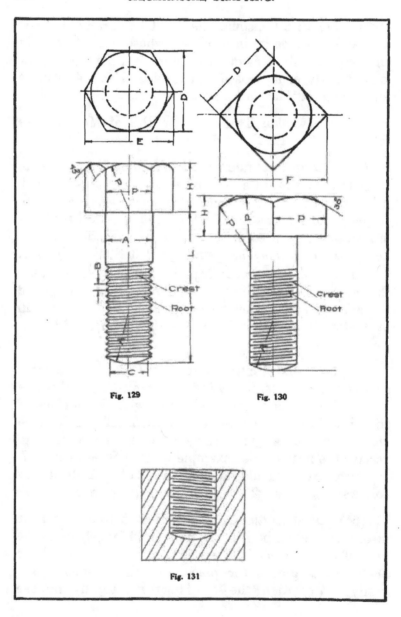

Fig. 129

Fig. 130

Fig. 131

cylinder and the head is drawn down with the regular standard nuts. See Fig. 128.

(65) **Bolt threads.** If represented exactly as it appears the ordinary U. S. S. thread would be shown as in Fig. 129. The *sharp edge* of the thread being known as the **crest** and the *division line* between the threads as the **valley** or **root**. The *large diameter* of the *thread* is, of course, the same as the *diameter* of the *bolt*. The *small* or *root diameter* will be found in the tables under the head of *tapping drill*.

(66) **Conventional threads.** To represent a bolt with the *notched edges,* as in Fig. 129, requires an unnecessary amount of care, and a conventional thread has been devised as a substitute; this is shown in Fig. 130. In this figure it will be seen that the notched edges are omitted. The *light lines* represent the *crest* of the threads and run entirely across the bolt, while the *short heavy lines* between these represent the *roots* or *valleys* of the threads, and are *limited* by lines whose *distance apart* is equal to the *distance C* or the *tapping drill* for that size bolt.

This same convention is used to represent the threads in a threaded hole, Fig. 131; however, it must be noted that the direction of slant of the thread lines is different; i. e., for the threads on the front of a bolt (*right-hand thread*) the lines slope from **left** to **right up,** while the lines for threads on the *back* of a bolt or a threaded hole slope from **right** to **left up.** Mistakes can easily be made in this; however, after inspection of an actual bolt it will be easily understood. In drawing the lines for conventional threads they should be given a slope of from 5 to 10 degrees, not more. The distance between the lines need not be scaled according to the pitch. The

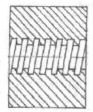

Fig. 132

Fig. 133

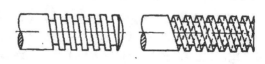

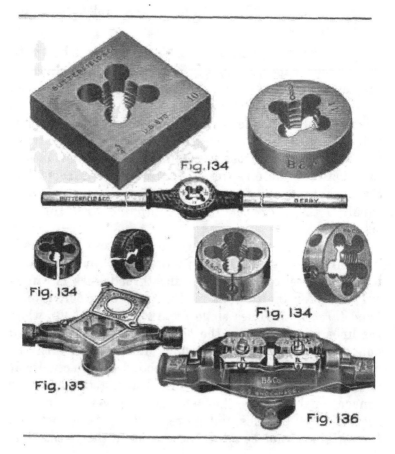

Fig. 134

Fig. 134

Fig. 134

Fig. 135

Fig. 136

number of threads per inch is invariably indicated by a note (e. g., ¾″x10 *pi*.) and the conventional lines may be spaced by the eye so as to appear symmetrical.

(67) **Square threads.** For bolts which are being constantly loosed and drawn, as in the tail stock of a lathe, etc., a thread is used in which the friction does not increase so rapidly with the tension as in the V thread. This is the **square thread,** the *bearing surfaces* of the threads being *perpendicular* to the *axis* of the bolt. No conventional representation of this thread has been devised, so they will always be represented as in Fig. 132.

(68) **Double square threads.** Whenever it is desired to thread a bolt or screw so that the nut or hand wheel will draw with the greatest speed, a *double* set of threads is used; i. e., *two* threads of the *same pitch*. Unless inspected closely it will appear to be only a single thread. This thread will be found on the screw which moves the center in the tail stock of a lathe. This thread is represented in Fig. 133.

BOLT AND PIPE DIES

(69) The thread cutting tools used in cutting threads on bolts and pipes are known as **dies.** The type of die used in cutting *bolt threads,* as shown in Fig 134, is made from a flat plate of steel thru which a hole is first drilled and tapped out with a *master tap;* then four holes are drilled around this center, giving four cutting edges. These dies have *no* taper, as the outside diameter of a bolt should be uniform. The dies used in cutting threads on pipe may be either *solid* or in *sections* as one individually prefers. Both types are shown in Figs. 135 and 136. All such dies have a standard taper of 1″ in 16″ so that any pipe threaded by one of these dies, when

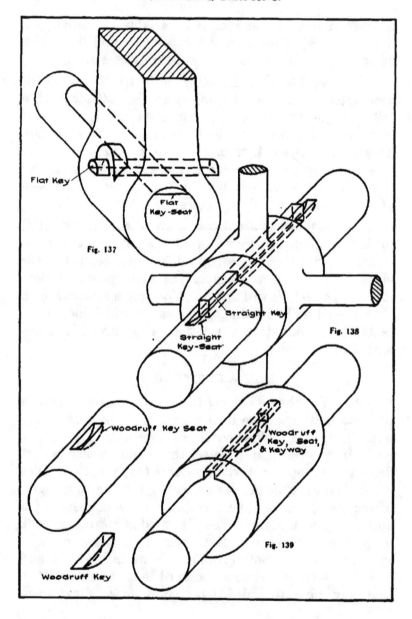

Flat Key

Flat
Key-Seat

Fig. 137

Straight Key

Straight
Key-Seat

Fig. 138

Woodruff Key Seat

Woodruff
Key, Seat,
& Keyway

Fig. 139

Woodruff Key

screwed into a pipe fitting, makes a water or steam tight joint when drawn sufficiently tight.

KEYS AND KEYWAYS

(70) In fastening wheels, pulleys, etc., to shafts two classes of fasteners may be used, *set-screws* or *keys*. Set-screws may be used to advantage in all cases in which the *twisting force* on the shaft is very *small*. It is not wise to use such a fastener on large pulleys or in cases in which the load is likely to be very great. In this latter case one of three varieties of key may be used according to the nature of the machine. These keys and keyways are (a) **Keys on flats**, Fig. 137; (b) **Straight seated keys**, Fig. 138; (c) **Woodruff keys**, Fig. 139. It is clearly seen that the first type of key has a very limited use, and owing to the fact that the Woodruff key is patented, the *straight keys* and seats have the most general use. In this type of key, it is seen, **Fig. 138**, that one-half of the recess is milled or cut into the shaft (this *groove* being known as a **keyseat**) while the other half of the recess is cut into the hub of the pulley and is known as the **keyway**. When fitted over each other they should form nearly a square, the height being slightly less than the width. The *keys* used in this connection are cut from *square bars* of *cold rolled steel* and filed to a slight taper on one side only, so that when driven in far enough they wedge between the hub and shaft, thereby preventing the pulley both from sliding along the shaft in either direction and from turning about the shaft.

CONVENTIONAL BREAKS

(71) In making a drawing of a shaft of uniform diameter it is quite frequently impossible to represent the

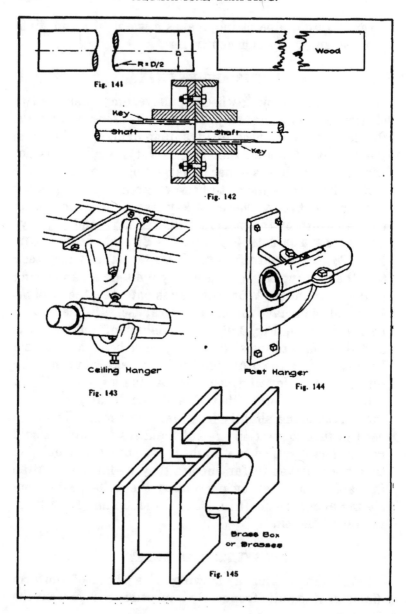

Fig. 141

Fig. 142

Ceiling Hanger

Fig. 143

Post Hanger

Fig. 144

Brass Box
or Brasses

Fig. 145

whole shaft without changing scale. As a substitute for this change of scale any convenient space may be used to represent the length of the shaft. A conventional break being shown at some place in this length, as in Fig. 140, the dimension given is for the *total length* of shaft. The same conventional break may also be used in a long tapered pin, the whole of which cannot be conveniently drawn. This form of break should be made mechanically while others, which may be necessary, can be drawn freehand, Fig. 141.

SHAFT COUPLER

(72) The shafting used in transmitting power thru a shop comes in standard lengths of from 20 to 40 feet. Any length of shaft may be made up from these standard lengths by various types of *shaft couplers,* one of which is shown in Fig. 142. One-half of this coupler is keyed to one end of one bar of shafting and the other half to the adjacent end of the next bar. The use of four or six bolts thru the webs of these coupler parts converts these pieces of shafting into one double length of shaft.

BEARINGS AND HANGERS

(73) **Hangers.** Lengths of shafts are supported by either of two types of *hanger,* the one known as a **wall hanger,** the other known as a **ceiling hanger.** In Figs. 143 and 144 are shown both types of hanger, the construction of which needs little explanation. The two half boxes or casings thru which the shaft passes are lined with an alloy of zinc and lead, which minimizes the friction and consequent loss of power in such bearings.

Brasses. In certain types of machines it is desirable to use brass for bearings rather than the combination of lead and zinc. Such bearings are ordinarily made in halves, constructed so as to permit of a slight adjustment, Fig. 145. These half bearings are commonly known as **brasses.**

Babbit. The alloy of lead and zinc mentioned above is commonly known as **babbit.** The greater the amount of zinc the harder this compound is. Two of the material advantages of babbit for bearings are, that the metal is cheap, and such bearings can be easily replaced by a workman of but ordinary experience. Babbit lining is poured while molten into the cast iron casings with the shaft in place.

LESSON 8

ASSEMBLY DRAWING

(74) **Definition.** An **assembly drawing** of a machine is a *two or three view orthographic projection of a machine completely assembled;* i. e. all parts in their proper working place.

Uses. Assembly drawings have three important uses. (1) As an index to a working drawing; i. e., an assembly of a machine is ordinarily given with the set of detail working drawings to explain the use of each of these details in the machine; (2) For purposes of *advertisement* or *magazine illustration;* (3) As a *construction guide* in assembling machines which may be sent out from shops in sections.

Characteristics. Some characteristics which may be noted of assembly drawings when used for any of the above purposes are: (a) All *dimensions* are ordinarily *omitted;* (b) The several views are *elaborately sectioned* to explain clearly all inside constructions; (c) When given with a set of details the assembly will ordinarily *occupy* a *fixed relative place on* the *sheet,* i. e., the *lower left corner* or *whole left side* if necessary.

DETAIL WORKING DRAWINGS

(75) **Arrangement of set of details.** In making a set of details a certain order of arrangement should be followed, both for appearances and ease in reading the drawing. As mentioned under assembly drawings, the assembly should occupy the left portion of the sheet. In gen-

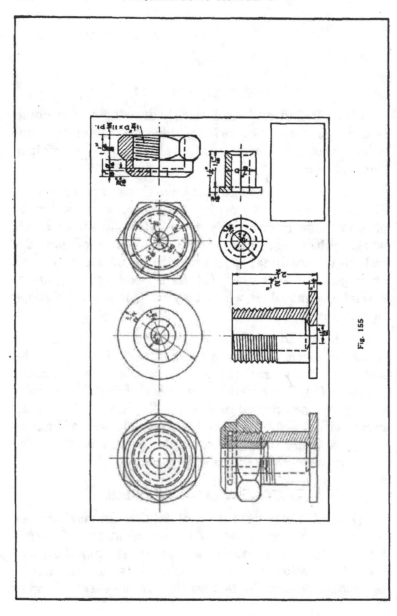

Fig. 155

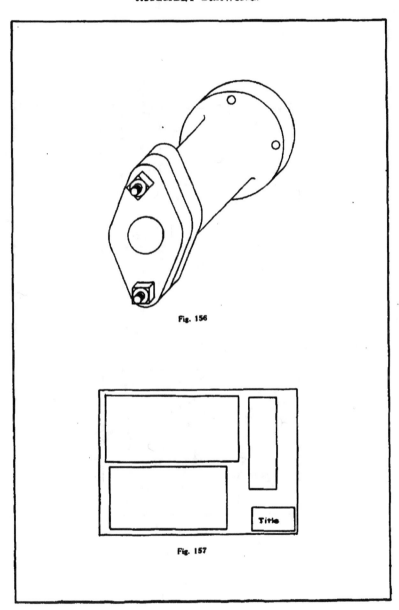

Fig. 156

Fig. 157

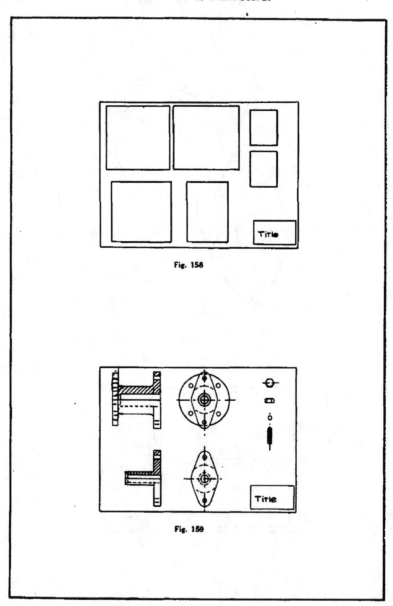

Fig. 158

Fig. 159

eral, the arrangement of the details on the sheet should be *such as to suggest their direct relation in the machine itself;* i. e., such an arrangement as is suggested by the relation of these parts in the assembly. It may not always be possible to carry out this scheme completely; however, in general it will be found possible so to arrange the main details. For further explanation see Fig. 155. The assembly is here shown to the left, followed along the bottom of the sheet by the main detail, the remaining details being arranged about the sheet properly with relation to each other if not to the main details.

Order of work in a set of details. In making a set of details, e. g., of the stuffing box, Fig. 156, the following order of work will be found to lend to the greatest speed: (a) Sketch roughly on a piece of scratch paper, rectangles for the various details in their proper arrangement, Fig. 157. (b) Enlarge upon the first rough sketch by placing on a second sheet of scrap paper rough rectangles for the necessary views of the details as they have been arranged on the first sheet, Fig. 158. (c) Block out roughly on the sheet on which the drawing is to be placed, using the scale only approximately, rectangles to correspond to the arrangement on sheet No. 8, Fig. 159. (d) Place in accurately the final rectangles by means of the scale and draw center lines of these rectangles for the center lines of the various views of the details, Fig. 159.

VALVES

(76) There are in general use at present valves of *two* distinct designs; most of us may be familiar with the appearance of both of these designs but hardly with the construction and uses. These two types are **Globe** and **Gate** valves.

Fig. 160

Copper Disc and Holder

Rubber Disc and Holder

Huxley Seat
Fig. 162

(77) **Globe valves.** In Fig. 160 is shown the common type of *Globe Valve,* so named from the shape of the main part of the body. In Fig. 161 is given a section of this valve showing clearly its construction and the name of each part. The steam, air, or water enters at the left, passes up thru the opening within the rim of the **seat** and on out to the right. When the **stem** is screwed down by the **hand wheel** the **disc** is wedged tight into the **seat,** thereby closing the opening and stopping the passage thru the valve. To prevent the pressure from forcing the contents out of the top of the valve around the stem, hemp or especially prepared **packing** is placed in the **bonnet,** about the stem and just under the **gland,** and the **gland nut** drawn down until packing is sufficiently tight to prevent escape.

Discs and seats. In Fig. 162 is shown the patented Huxley seat now used in Nelson globe valves, also various types of discs used for various purposes. The reason for the use of the Huxley seat is that the grit that may be carried thru the pipes by water and steam rapidly cuts out the seat, making it necessary to regrind it frequently to keep the valve in perfect condition. These patented seats being made of copper, which is comparatively soft, keep in good shape for quite a time from the occasional pressure of the disc, and when worn badly may be easily replaced. Valve disc No. 1 is of *solid brass* and is quite common on valves. Disc No. 2 is of *brass* and has a recess in the bottom into which may be fitted a *hard rubber* or *lead* disc for use on *air* or *water* lines. Disc holder No. 3 is fitted with a *copper ring* or *disc* for use on *steam* lines.

Bonnets. In Fig. 163 is shown the **coupler** type of bonnet which screws directly to the body. To regrind

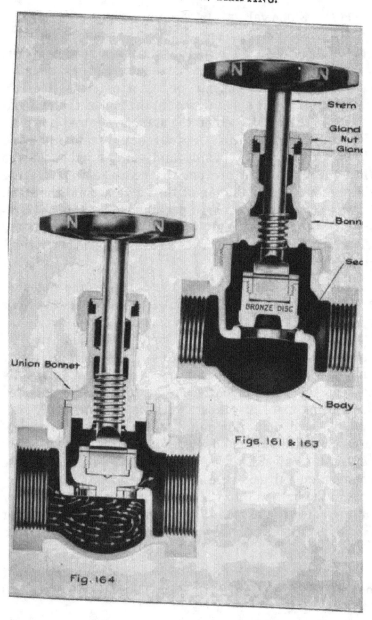

Fig. 164

Figs. 161 & 163

the valve seat it is necessary to remove this bonnet; hence, to prevent sticking of the bonnet and trouble in removing, no packing is used at the point indicated; a brass tight joint is depended upon. However, in spite of good workmanship, water and steam may force limestone and dirt out under the bonnet, causing it to stick quite badly. If much force is necessary to remove the bonnet the body may be badly twisted and the valve ruined. To eliminate these troubles the **union** type of bonnet, shown in Fig. 164, has been designed and finds great favor. In this type the bonnet does not turn as the union nut is run down, so that packing may be used if desired without the danger of bad sticking.

Angle valve. In Fig. 165 is shown a type of globe valve known from its design as an **angle** valve; the water or steam enters and leaves as indicated. The internal construction of the valve is similar to that of the straight type of globe valve.

Objection to globe valves. It is easily understood that especially in water lines any *obstruction* which may be placed within the pipe has a tendency to reduce the *pressure* of the water when in motion, and the pressure is usually an item of considerable importance. Any *angular turns* about which the water must move reduce the pressure by *friction*, and *direct obstructions* reduce the pressure by *back currents*. As the water passes thru the *globe valve*, Fig. 164, pressure is reduced in both of these ways; the water must make two right angular turns in passing thru the valve; furthermore, as it strikes the diaphragm a return current is created, subtracting just so much from the pressure. For these reasons the *gate valve*, which permits a *straight* passage of fluid when open, is considerably more efficient for water lines.

Fig. 165

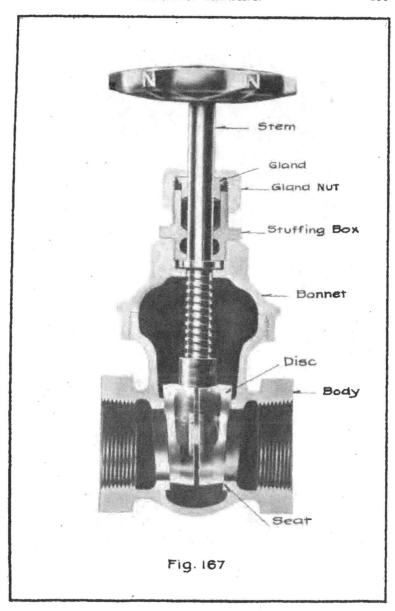

Fig. 167

Fig. 169

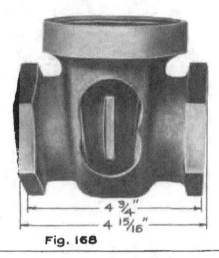

Fig. 168

Fig. 170

Fig. 168

Fig. 171

Fig. 172.

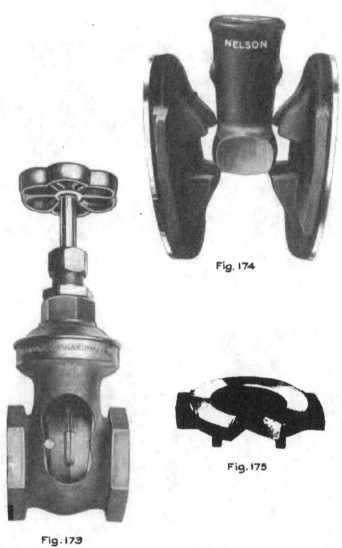

Fig. 174

Fig. 175

Fig. 173

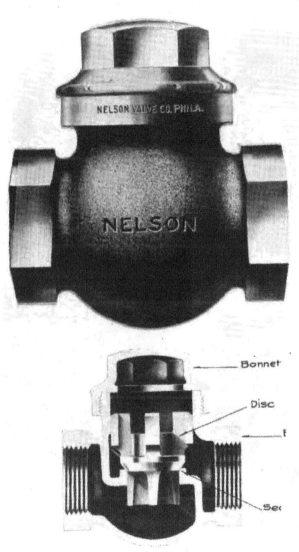

Bonnet

Disc

Se

Fig. 176

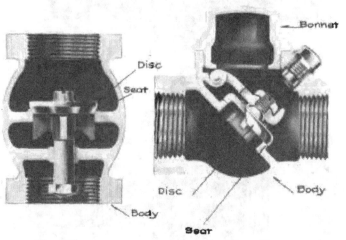

Fig. 177   Fig. 178

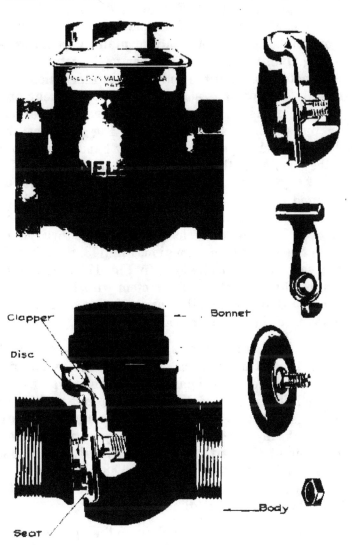

Clapper

Disc

Bonnet

Seat

Body

Fig. 179

(78) **Gate valves.** In Fig. 167 is shown the construction and names of the various parts of the gate valve. The original design of gate valve contained a **solid wedge** disc, Fig. 168, which proved under various tests rather unsatisfactory; e. g., if the body were strained, this strain ordinarily developed as shown in Fig. 169, thereby opening up the seats and ruining the valve. Furthermore, when put to the test as shown in Fig. 170, the *solid wedged* disc being pushed in place by hand, it was found that the opening in the seat was by no means closed. The **double wedge** type was then designed as a substitute and was found to stand all the above tests satisfactorily, as shown in Figs. 171 and 172. It was also found that even when it is impossible to close the opening thru the one seat, on account of an obstruction, Fig. 173, the other side of the valve closes perfectly. In Fig. 174 is shown one design of the connection of the stem with the two discs, and in Fig. 175 is shown the *copper facing* of a disc as it is rolled into the cast iron body.

(79) **Drop check valves.** In Fig. 176 is shown the common type of **drop check** valve with the construction and names of the various parts. As steam or water enters from the left it raises the disc and passes out to the right. As soon as the pressure is relieved from the left the disc drops back in place, preventing any return passage. Fig. 177 is of the **vertical drop check** valve to be used on any vertical section of pipe. The *objection* to the valve shown in Fig. 176 is the same as that against *globe valves,* Fig. 166, i. e., it diminishes the *pressure.*

Swing check valves. To eliminate the objection to the *drop check* valve the **swing check** has been designed as shown in Figs. 178 and 179. Type (*a*) is perhaps prefer-

ASSEMBLY DRAWING.

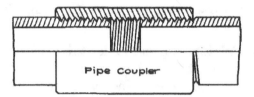

Pipe Coupler

Fig. 180

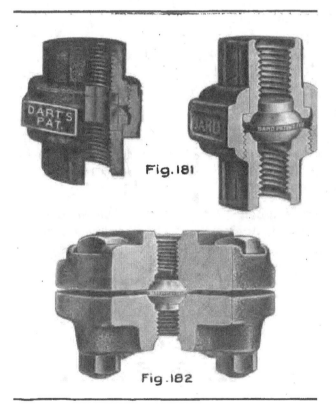

Fig. 181

Fig. 182

able, as it presents a more perfect passage for the water than type (*b*).

(80) **Water pipe and boiler tubes.** In speaking of a 2″ *water pipe* it should be understood that the **inside** diameter of the pipe is indicated, the *inside* diameter being necessary to compute the *quantity* of *fluid* passing thru the pipe. On the other hand a 2″ *boiler tube* is a tube whose **outside** diameter is 2″, the *outside* diameter in this case being necessary to compute the *heating surface* and *horse power* of a boiler.

PIPE CONNECTIONS

(81) **Pipe coupler.** In Fig. 180 is shown a pipe connection known as a **coupler**, used in connecting lengths of pipe into a line.

Ground joint union. In Fig. 181 is shown a pipe connection known as a **union**, to be used in a pipe line wherever it is likely that the pipe may have to be uncoupled for repairs, etc.

Flange union. In Fig. 182 is shown a connection known as a **flange union** which is used as a substitute for the union mentioned above on most all pipes of diameter greater than 2½″.

LESSON 9

ISOMETRIC PROJECTION

(82) **Definition and explanation of principles.** Tho the fundamental principles depend upon orthographic projection, they are so easily understood that it will be possible for the student to grasp them fully even with a limited knowledge of orthographic projection. **Isometric projection,** as the term indicates, *is a projection of equal or proportional measurements.* If thru a given point called an **origin,** *three lines* be drawn at *right angles* to each other, e. g., the *three adjacent edges* of a *cube,* we have the *three coordinate axes, x, y,* and *z* known in analytic geometry. A *fourth line* passed thru the given point and at equal angles with the first three lines is known as the **Isometric Axis;** it may be compared to the *diagonal* of a *cube.*

A *plane perpendicular* to this *axis* is the **isometric projection plane;** for, since the *coordinate axes* make *equal angles* with the *isometric axis* they must make *equal angles* with this *projection plane;* and *equal lengths* on the *coordinate axes* or on *lines parallel* to the *axes* will project as *lines of equal length* on this *plane.*

DIRECTRICES

(83) The *orthographic projections* on this *plane* of the *coordinate axes* are known on the drawing as the **directrices,** and occasionally as the **isometric axes.** Since the coordinate axes make with each other equal angles, their projections *(the directrices)* also make equal angles (120 degrees) with each other. This being true, when it

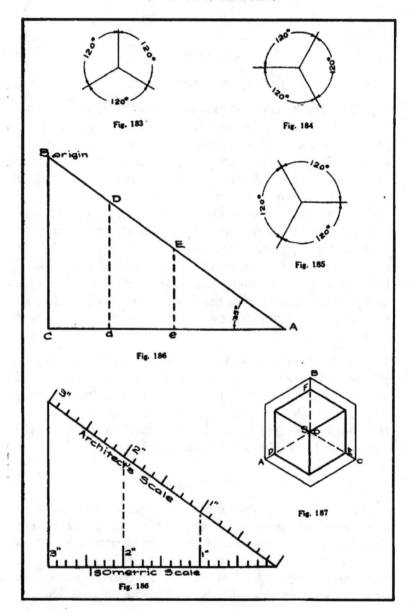

Fig. 183

Fig. 184

Fig. 185

Fig. 186

Fig. 187

Fig. 186

is desired to make any isometric projection the directrices may be drawn immediately thru a chosen point and at 120 degrees to each other. One of the directrices is usually taken *vertical*, Fig. 183; however, the arrangements shown in Fig. 184 and 185 are frequently used.

ISOMETRIC SCALE

(84) Since the coordinate axes are oblique to the isometric plane their projections are shorter than the axes themselves. Each coordinate axis makes with this isometric plane an angle of about 35 degrees. Then assuming the hypotenuse, AB, of the right triangle, Fig. 186, to be one of the *coordinate axes*, BC, the *isometric axis*, and CA, the *isometric plane*, viewed *edgewise*, the *isometric projection* of any given length, DE, on the axis BA, would project on the plane with length equal to *de*.

To make the isometric projection of proper proportions, divide the hypotenuse of a triangle similar to ABC into inches, etc., and project these inches upon CA. This scale obtained on CA is known as the **isometric scale** and lines *parallel* to the *directrices* should be measured according to it instead of to the architect's scale.

Problem 1. **To construct the isometric projection of any parallelopiped.**

Construction. Cube 2″ on edge, Fig. 187.

Thru point O are drawn the axes OB, OA, and OC, making with each other angles of 120 degrees. From O, along OA, measure with the isometric scale 2″; the same along OC and OB to points D, E, and F. OD and OE then represent the two adjacent sides of the base, and lines from D and E parallel to OE and OD complete the base. In similar manner the vertical faces DOF and FOE

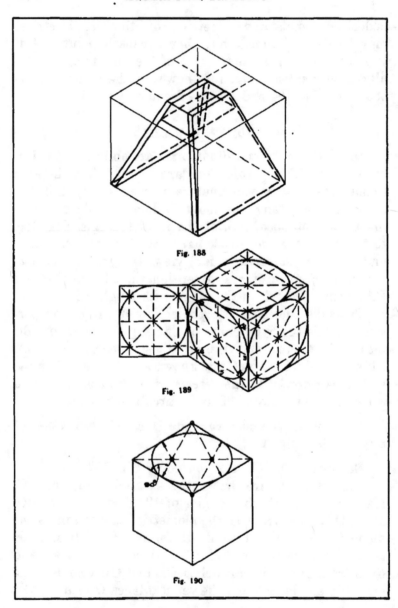

Fig. 188

Fig. 189

Fig. 190

are completed; then draw the top base FS and finally the faces SD and SE.

IRREGULAR OBJECTS

(85) Inasmuch as all but a very small percentage of machine parts are either of rectangular shape or can easily be enclosed in such a box or parallelopiped, the isometric projections of irregular objects are easily constructed with the aid of such an enclosing parallelopiped, Fig. 188.

CIRCLES

(86) Problem 2. **To construct the isometric projection of a circle.**

8 POINT METHOD

Construction. ABCD is the isometric projection of a square in which a circle is inscribed. On edge AB construct a square and inscribe in it a circle. Draw the horizontal diameter of this circle and diagonals of the square. The points of tangency of the circle with the square and the 4 points of intersection with the diagonals constitute the desired 8 points. If the diagonals of the parallelogram be drawn the isometric projections of these 8 points may be obtained as shown and the ellipse drawn thru them freehand. Fig. 189.

APPROXIMATE MECHANICAL METHOD

Fig. 190 illustrates an approximate mechanical method for obtaining the isometric projection of a circle.

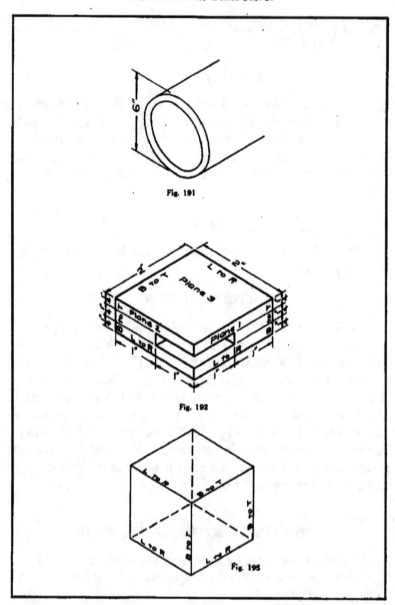

Fig. 191

Fig. 192

Fig. 195

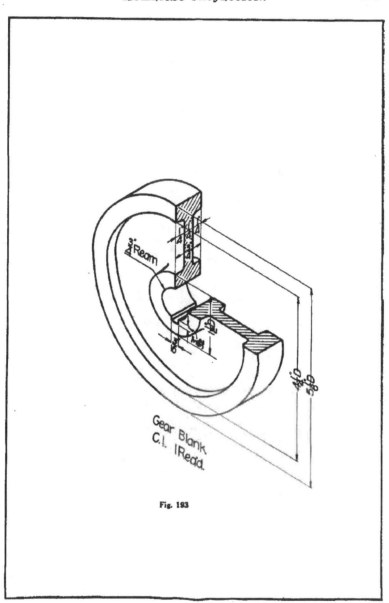

Fig. 193

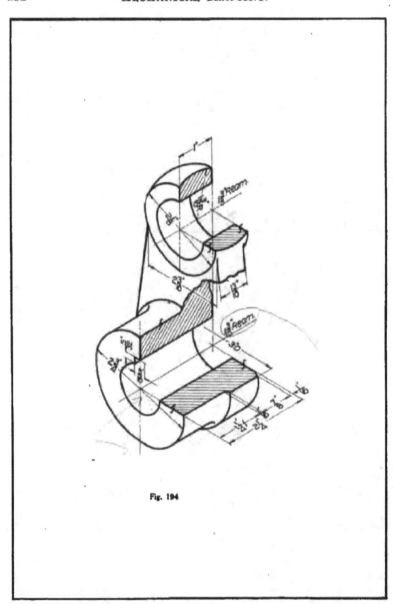

Fig. 194

DIMENSIONING

(87) In placing dimensions on an isometric drawing, the same rule must be followed as in orthographic working drawings; the dimensions must read from *left* to *right* or from the *bottom up*. In following this rule it will be found always that the dimension lines are parallel to the coordinants axis, **never** otherwise. In giving the *diameters* of *circles* the method shown in Fig. 191 is preferable to placing the diameter directly on the isometric of the circle. On inspection of Fig. 192 it is seen that there are actually three faces of the object to be dimensioned and in giving the dimensions for face No. 1, which is parallel to the isometric plane No. 1, it must first be decided what directions constitute from *left* to *right* and from *bottom up*. The same thing must be decided for faces 2 and 3. In Fig. 195 is given a key which will be useful in dimensioning. In connection with this key it will be necessary for the student, in placing dimensions, to decide merely to which face of this key his dimensions are parallel, Figs. 193, 194.

SHADES AND SHADOWS IN ISOMETRIC OBLIQUE PROJECTION

(88) Shades and shadows, in isometric projection as in orthographic projection, are used merely for the natural appearance which they give to a drawing. Hence, as in orthographic projection, the draftsman has considerable freedom in determining both the *direction* and the *length* of the shadows. Before proceeding with a problem in shades and shadows it will perhaps be well to define a few of the terms and state the fundamental principles which govern construction.

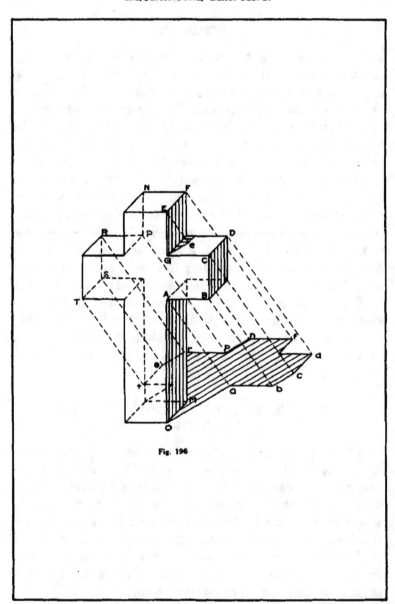

Fig. 196

Direct light. All light which comes to any body directly from the source, the sun, arc lights, etc., is known as **direct** light.

Indirect light. All light which reaches objects in an indirect way, e. g., by reflection from other objects which are in direct light, is known as **indirect** light.

Shade. Any portion of the surface of an object from which the *direct light* is excluded by some part of the *same* object is said to be **in shade.** This shaded surface may also be known as **a shade.**

Shadow. Any portion of the surface of an object from which the *direct light* is excluded by some part of *another* object is said to be **in shadow,** and for convenience may be called a **shadow.**

PRINCIPLES

1. All rays of light in these problems are regarded as *parallel;* hence, when the direction of the first ray has been assumed, all others must be considered parallel to it.

2. The shade or shadow of a given point on a given surface is the point in which a ray thru the given point pierces the given surface.

3. Any *ray* used to determine the *shade* or *shadow* of a given point may in reality be a ray of light; however, in this discussion it will be known as a **shadow ray;** see Miller's *Descriptive Geometry,* Art. 124.

4. If a line AB is parallel to a plane, e. g., H, the shadow of AB on H is *parallel* and equal in length to AB.

5. If lines AB and CD are *parallel* the *shadows* of AB and CD on any plane must be *parallel,* i. e., the *shadows* of *parallel lines* are *parallel.*

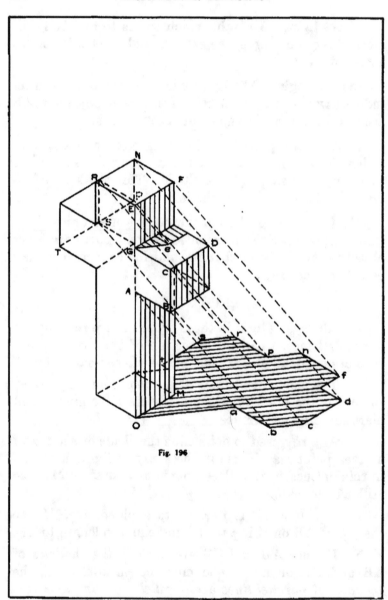

Fig. 196

6. If a line AB is oblique to H, AB and its shadow on H will meet at the point in which AB pierces H.

7. The *shadows* of *parallel lines* on *parallel planes* are *parallel*.

Problem. **To find the isometric or oblique of the shade and shadow on H of a given object.**

Given. Isometric and oblique projections of Cross, Fig. 196.

Req'd. Isometric and oblique projections of shade and shadow on H.

Beginning with point O as an origin, a line may be drawn in any desired direction, e. g., Oa, to represent the shadow of OA, and point a assumed in any desired position as the shadow of point A. Aa is the isometric projection of the shadow ray thru A, and all other shadow rays must be *parallel* to Aa. Since AB is parallel to H, its shadow ab is parallel and equal in length to AB. BC is parallel to OA, hence its shadow bc is parallel to Oa, and c is determined by shadow ray Cc. CD is parallel to H, hence its shadow cd is parallel and equal in length to CD; d may likewise be located by the shadow ray Dd. Since the plane GCDK is parallel to H, the shade of GE on this plane is parallel to Oa, see Principle 7, and e is located by the shadow ray from E. EF is parallel to GCDK, hence the shade line from e is parallel to EF; if the arm of the cross were not in the way, the shadow of E would fall at e_1; EF is parallel to H, hence e_1f is parallel and equal in length to EF; fn is parallel and equal in length to FN. It is readily seen from the direction of Oa that the face AOM must be in shade, likewise BCD and GEFK and space KGe-.

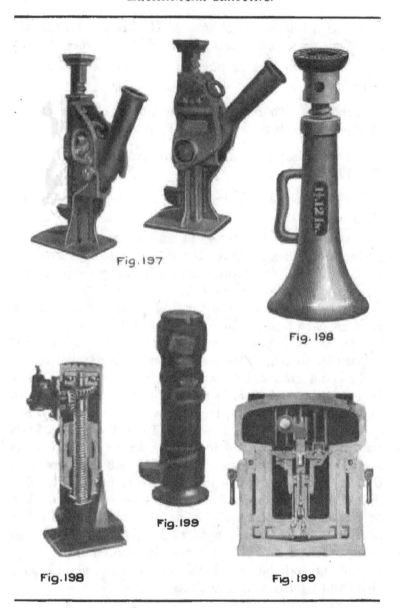

Fig. 197

Fig. 198

Fig. 199

Fig. 198

Fig. 199

The isometric projections of the shadows of the remaining points are located successively as those already found.

LIFTING JACKS

(89) **Lever jack.** In Fig. 197 is shown the **lever** type of jack used in lifting moderate loads rapidly. The speed with which this jack can be used is the main point in its favor.

Screw jacks. In lifting excessive loads the common jack used is the **screw**. The *screw* is turned by a bar inserted in the hole of the *capstan* head, running the screw either up or down. Fig. 198.

Hydraulic jack. Where extreme loads are to be lifted the **hydraulic jack,** Fig. 199, will be found most useful. The jack is composed of two cylinders and two pistons, the larger piston being forced up by the pressure of the fluid pumped into the cylinder by the smaller piston. Alcohol or oil may be substituted for water if the jack is to be used in cold climates.

CONSTRUCTION OF GEARS

(90) **Cog wheels,** pinions, etc., are first cast *blank,* as shown in Fig. 193, and the solid rim later cut into cogs on a milling machine.

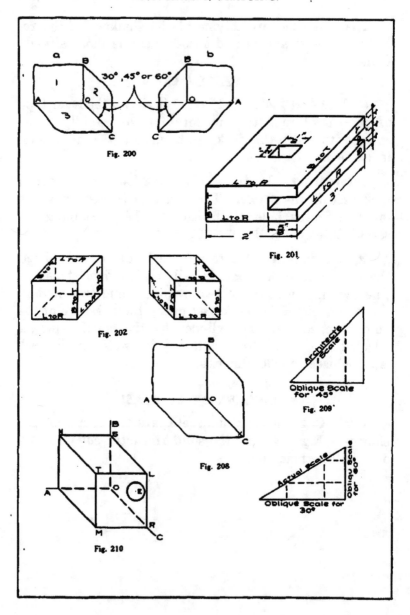

Fig. 200

Fig. 201

Fig. 202

Fig. 208

Fig. 209

Fig. 210

LESSON 10

OBLIQUE PROJECTION

(91)   Inasmuch as circles are so rarely found in such positions that their projections are true circles in *isometric projection,* a variety of projection has been devised in which it is possible so to place circles that their *projections* are *circles* and are easily drawn. This is known as **oblique projection.**

Inasmuch as the principles of oblique projection depend upon the theory of perspective, it will perhaps be better not to attempt any explanation of principles. Likewise, it may be well to mention that oblique projection is almost entirely a combination of incorrect principles, tolerated merely because of the ease with which this projection can be handled. The *directions* of the three directrices or oblique axes are correct; it is true, also, that when placed in one of the coordinate planes the oblique projection of a circle is a *circle;* with these exceptions the theories are incorrect. The principles on which oblique projections are made are as follows:

AXES AND CO-ORDIDNATE PLANES

(92)   Thru the origin, O, Fig. 200, a and b, are drawn three directrices or axes as shown; one *horizontal,* a second *vertical,* and the third to the *right* or *left* at 45°, 30°, or 60° with the horizontal, preferably 45 degrees. These three lines represent lines at *right angles* to each other, and may be compared to the *three adjacent edges* of a *cube.*

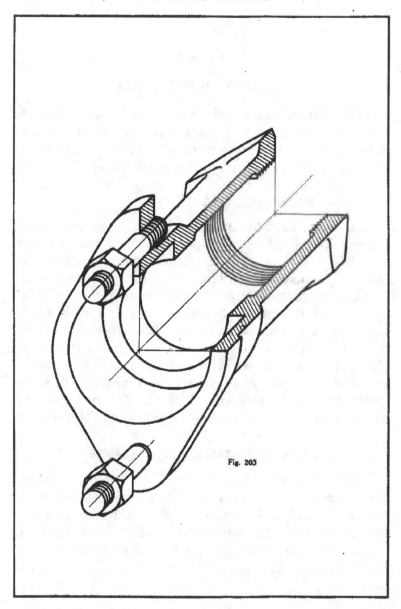

Fig. 203

Taken two and two, these axes include coordinate planes as follows: OA and OB, plane 1, or the plane of *true circles;* OB and OC, plane 2, a second vertical plane, and OA and OC, plane 3, a horizontal plane.

DIMENSIONING

(93)   In oblique as well as in isometric projection, the problems of dimensioning are threefold; i. e., any object drawn may have faces parallel to each of the three coordinate planes in Fig. 201.   In placing dimensions for constructions on these faces it is necessary, of course, to decide what directions constitute from *left* to *right* and from *bottom up.*   The key in Fig. 202 may be used in a manner similar to that of the key given for isometric projection.   Inspection of Figs. 203-207 may serve to clear up any doubtful points, both in dimensioning and in construction.

OBLIQUE SCALE

(94)   If *equal distances* be measured from O along the axes OA, OB and OC, Fig. 208, the distance along OC will appear to be *longer* than those along OA and OB; hence, for symmetry, it is necessary to make use of the oblique scale in measuring any distance along OC. The oblique scale is obtained by measuring off inches, etc., on the hypotenuse of a 45°, 30°, or 60° triangle and *projecting* these inches upon *either* of the *legs* for 45°, *long leg* for 30° and *short* leg for 60° (Fig. 209).

Problem 4.   **To draw the oblique projection of a parallelopiped.**

Construction. Cube 1″ on edge.

Thru a chosen point, O, Fig. 210, draw the three axes

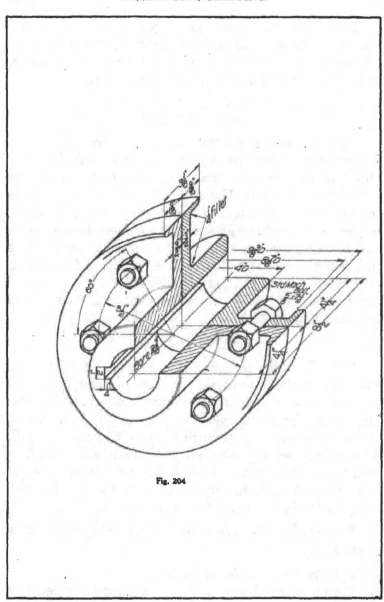

Fig. 204

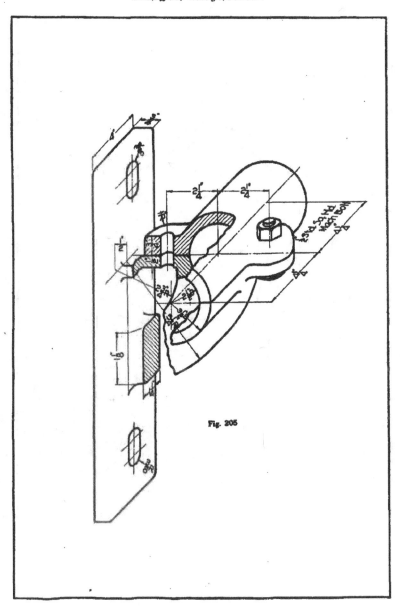

Fig. 205

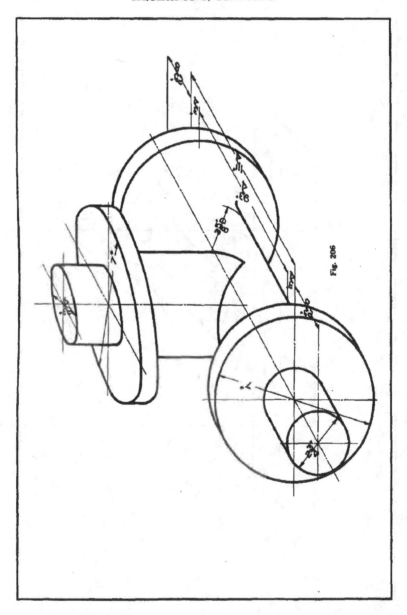

Fig. 206

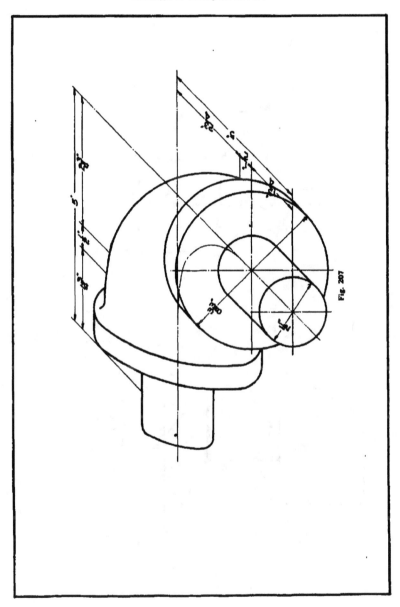

Fig. 207

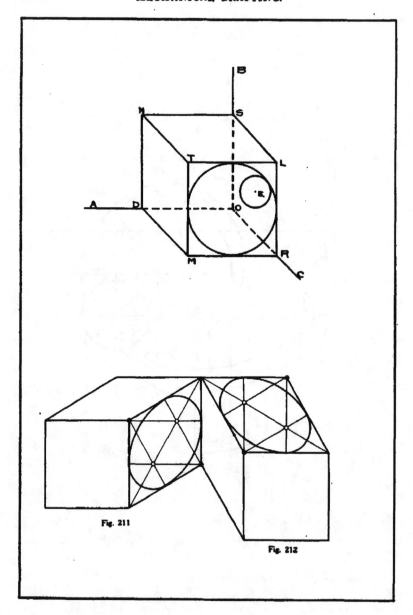

Fig. 211

Fig. 212

OA, OB, and OC. Along axes OA and OB measure 1″
to D and S, and with the oblique scale measure 1″ along
OC to R. Complete the parallelograms ODMR, OSLR,
and DNSO; then the remaining faces may easily be
added.

CIRCLES

(95) Point E, in face TLRM, is the center of a circle
¾″ diameter. Since this is the face of *true circles,* the
circle may be constructed with the compass. If it is
desired to draw the projections of circles lying in faces
TNSL or NTMD the *8-point* method explained in isometric projection may be used.

If the oblique axis be drawn at an angle of *30 degrees*
to the *horizontal* it will be noted, in Fig. 211, that face
NM is now of such shape as to convert the oblique projection of any circle placed in that face into an *isometric
projection,* and the *mechanical method* of constructing
that isometric projection, shown in Fig. 190, can and
should be used. If, on the other hand, the *oblique axis*
be drawn at an angle of *60 degrees* to the *horizontal,*
Fig. 212, the face NL is now of such shape as to convert
the *oblique projection* of any circle placed in it into an
isometric projection which can be constructed *mechanically* by the method of Fig. 190.

With some care it will now usually be possible, in
making the oblique projection of any object containing
circles, so to place the object that the oblique projections
of some of the circles are *true circles,* while the projections of the remainder become *isometric* when the *oblique axis* is drawn either at *30* or *60 degrees* to the
horizontal.

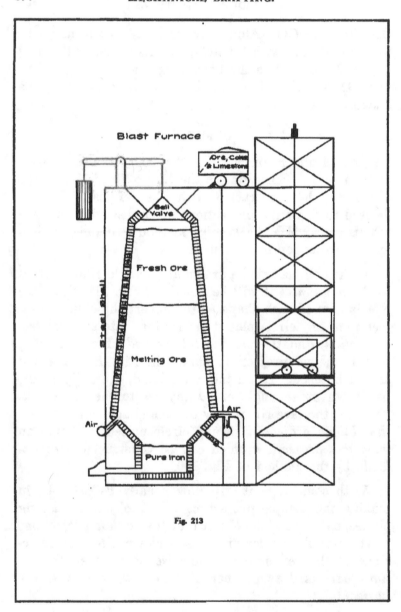

Fig. 213

IRREGULAR OBJECTS

(96)   All that has been said on these subjects under isometric projection applies as well in oblique projection. It is perhaps unnecessary to suggest that when making any oblique projection care should be taken so to place the object that most of the circles will appear as *true* circles.

CONVERSION OF IRON ORES INTO COMMERCIAL IRON AND STEEL

(97)   **Ores.** The ores from which **pig iron** is made are *three* in number: 1. **Iron carbonate,** a compound of *iron* and *carbon,* also known as **specular iron,** from which about 1% of the commercial iron in use is made. 2. **Magnetite,** a compound of *iron* and *oxygen,* known as *magnetite* because of its slight *magnetic properties,* from which about 13% of the iron in use is made. 3. **Hematite,** also an *oxide* of *iron,* and of two varieties, *red* and *brown,* from which the remaining 86% of commercial iron is made. *Hematite* is a heavy reddish brown ore which is found in various localities over the country, two very important deposits being the Lake Superior, of Michigan and Wisconsin, and the Alabama, about Birmingham. In some localities the ore is found near the surface, and is so weathered or rotted that it can be scooped out by steam shovels; this is the case in the Lake Superior regions. In other localities the ore is so hard and in such masses that it is necessary to blast it out.

Conversion of iron ore into pig iron. The iron ore as taken from the mines is hauled to the furnaces and crushed to egg or walnut size. In converting the iron ore into pig iron it is necessary to take from the ore the

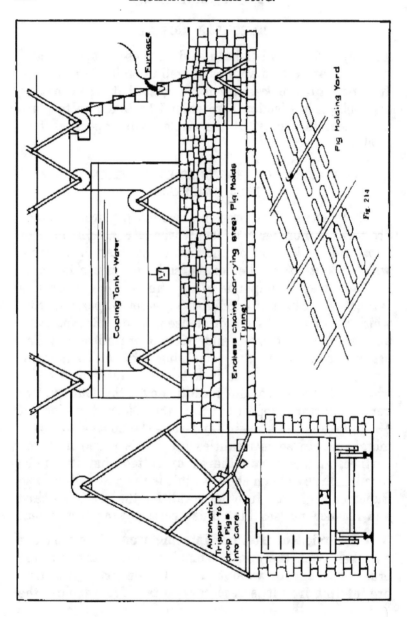

Furnace

Cooling Tank~Water

Endless chains carrying steel Pig Molds

Tunnel

Automatic Tripper to drop Pigs into cars.

Pig Molding Yard

Fig 214

oxygen and combine with the iron in place of the oxygen a small percentage of *carbon.* To accomplish this, *coke* and *crushed limestone* are mixed with the ore and fed into the furnace, a section of which is shown in Fig. 213.

The *coke* is principally *carbon*, and the *limestone* a compound of *calcium* or *lime* and *silica* or *glass.* As the charges of ore, coke, and limestone are fed into the top of the furnace, the large bell valve at the top opens from the weight, closing immediately after the charge is in, thereby preventing the escape of gases thru the top of the furnace. This charging of the furnace is carried on at regular intervals both day and night. It takes about 36 hours for the ore to drop from the top of the furnace to the bottom and be drawn off as pig iron. As it gradually drops, the heat generated by the burning of part of the coke by the heavy blasts of air forced in at points indicated, heats it to the melting point, where *carbon* from the *coke combines* with the *oxygen* of the ore, making two gases, **carbon monoxide** and **carbon dioxide**, which work their way to the top of the furnace and are drawn off thru a pipe at the side. As the *carbon* takes the *oxygen* from the *ore,* a small percentage of *carbon* combines with the *iron* in the place of the *oxygen,* and the *dirt* of the ore, *clay,* etc., combines with the *melted limestone* to form what is commonly known as **slag**. The pure iron drops to the bottom of the furnace, to the hearth, and the melted slag floats on this melted iron. At regular intervals the slag is drawn off in cars and hauled to dumps, and the iron drawn off and molded into **pigs**. Two methods are used in the molding of this iron, the one by converting a considerable space of ground, covered with sand, in front of the furnace, into one large mold. A large trough leads from the furnace

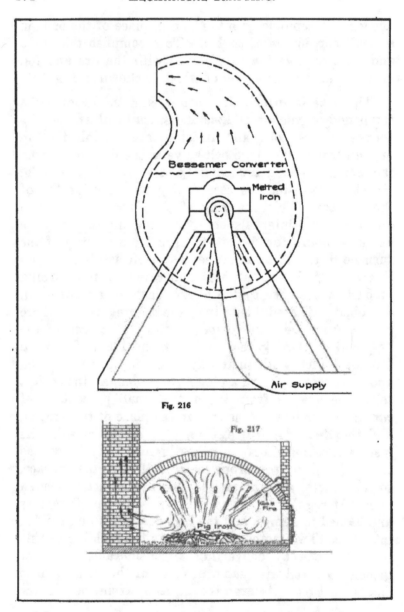

Fig. 216

Fig. 217

down thru the center of the yard; smaller lateral troughs
lead from this large trough, and individual troughs feed
into the depressions where the pigs are to be cast, Fig.
214. The second method is by a molding machine consisting of two endless chains carrying between them
soft steel molds, into which the pig iron is poured at the
furnace. These chains, moving slowly, pass the molds
of iron thru a tank of water, cooling them, and on to a
point where they are automatically unloaded and the pigs
of iron dropped into cars, Fig. 215.

Pig iron is rather soft, and when melted in cupolas
of a foundry, to be run into castings, it is usually necessary to add a certain quantity of scrap cast iron to make
the metal harder.

Bessemer steel. To convert pig iron into one of the
classes of steel used in making railroad rails and standard construction iron, the melted iron is hauled in cars
from the blast furnace to what is known as the *Bessemer
converter*, Fig. 216. This converter is supported on two
pinions or trunions, and can be tipped over to pour the
charge in or out. After the charge of melted iron is
poured into the converter and the converter shifted upright, air at about twenty pounds pressure and heated
to about 800° F. is forced into the converter as indicated,
causing the iron to boil most violently. This air both
burns out some of the *carbon* and *blows* out any *slag*
which may be left in the iron. This boiling is continued
from ten to twenty minutes and discontinued when the
flame from the mouth of the converter takes on a certain color. The converter is then tipped over and a quantity of the compound of *iron* and *nickel*, known as
Spiegeleisen, thrown in. After boiling for a minute or
two more, to mix this nickel thoroly thru the mass, the

iron is cast into large billets, which are later rolled into
steel rails, etc. This steel coming from the converter is
known as **Bessemer Nickel Steel,** Bessemer Manganese
Steel, etc., according to the alloy which was added to give
it greater strength.

Wrought iron. To make **wrought iron,** pig iron is
broken into pieces and placed on the slag-covered hearth
of what is known as the **reverberating furnace,** Fig. 217,
so called from the fact that the top is so shaped as to
throw the gas flame directly down on the mass of iron.
As the pig iron becomes plastic it is stirred by rods,
worked thru holes in doors about the furnace, until most
of the slag that remains in the iron when poured from
the blast furnace has been worked out; it is then allowed
to cool slightly and taken out and rolled into a sheet.
After several of these sheets have been rolled they are
heated and welded together and rolled into the wrought
iron bars of commerce. The process of working the slag
out of the iron in the reverberating furnace is called
puddling.

Crucible or tool steel. Crucible steel, which is used
for all varieties of cutting tools, is made from *wrought
iron.* Bars of wrought iron are cut into pieces and
about 100 lbs. of these pieces, with perhaps 1 lb. of charcoal, is placed in a covered earthen crucible and heated
in a furnace until the iron has been melted and has absorbed all of the carbon. This iron is then cast into
small billets and later rolled into convenient bars for
tools. The introduction of the carbon into the wrought
iron has given to the resulting compound the property of
extreme *hardness.*

Open hearth steel. A second class of steel, known as **open hearth steel,** is made in a furnace similar to the reverberating furnace used in manufacturing wrought iron. This steel is made, however, by working into the melted pig iron a quantity of scrap iron, producing in general a steel of a better quality than the Bessemer. This class of steel can be used for *steel rails, I beams,* and in *sheets* for the making of *boilers,* etc. The Bessemer steel mentioned above is ordinarily used only for *rails* and *construction iron,* and **not** for *boiler iron.*

LESSON 11

MACHINE SKETCHING

(98) **Definition.** A **machine sketch** may be roughly defined as a *freehand working drawing*.

To the engineer no one accomplishment is of more value than the ability to make rapidly accurate, legible machine sketches.

A draftsman or shop foreman may be called upon at any time to make a hasty sketch of some broken machine part which perhaps cannot be removed without shutting down the machine for a day or two.

A construction engineer putting in some new machinery may find that some plates, fixtures, etc., designed especially for the job, are all wrong, and he must immediately send in sketches of what is wanted.

A bridge engineer may find his work held up by the breaking or absence of some peculiarly shaped piece, or may need some special fixtures to handle difficulties peculiar to the job.

Likewise, it is understood that all machine forms are devised in the mechanic's brain and must be placed on paper in some approximate form before it is possible to make a mechanical working drawing.

In all of these and hundreds of other cases which are inevitable, the ability to sketch *rapidly* and *well* is indispensable, and the man who finds himself called upon to make a sketch and is not well grounded in its principles, will find himself seriously handicapped.

(99) **Paper.** It will be seen that the nature of the situations which require sketching will demand the use

of scratch paper or a notebook. *Cross-section paper* is invaluable for this purpose, as it aids materially in the rapid and accurate sketching of the several views.

(100) **Nature of drawing.** As in a *mechanical working drawing*, a *machine sketch* consists of a number of views (*top* and *front; top, front,* and *left end,* etc.) of a machine or machine part. These views are true orthographic projections, hence projections of each other, as in working drawings. Never show more views than are necessary to explain clearly the construction; of course, *two* are a *minimum;* variations in this respect will be mentioned later.

(101) **Pencil—Sketch stroke.** For sketching it will be better to use a comparatively *soft* pencil, H or 2H, as it is desirable to show marked distinction between the outlines of the *object* and *dimension* and *section* lines.

In drawing lines, whether short or long, the *sketch stroke* should be used. The **sketch stroke** is merely a *succession* of *short strokes* in the desired direction, and, as a result, the line will, of course, be somewhat ragged, consisting of a number of short overlapping lines. However, by this method it will be found possible to approximate a straight line much more closely than by a continuous stroke.

(102) **Size of drawing.** The work being freehand and done usually under adverse conditions, sketches are not made to scale; *numerical dimensions* are depended upon entirely for sizes. As an aid in approximating proportions of the different parts of a machine, the following scheme will be found useful: Suppose after careful inspection it is decided that only two views are necessary, and these *front* and *right end.* You have perhaps a 5 x 7 notebook at hand and must place these

two views, with dimensions, on this size sheet. Estimate the ratio of the *length* of the object to its *width* and height and block out roughly on the sheet the proper proportional spaces, for the two views, making them as large as possible. Then measure off on a lead pencil with the thumb-nail a distance equal to the length you have given the space for the front view, and, holding the pencil horizontally and about 1′ from the eye, move off from the machine until the space from the end of the pencil to the thumb-nail just covers the length of the machine. Standing in this position and using the pencil in this manner, the several parts of the machine may be rapidly sketched in in their proper sizes.

(103) **Procedure.** In making a machine sketch, the greatest speed and accuracy will be attained by following some system. The following will be found valuable:

1. Decide on the *number* of *views necessary,* and decide which these should be.

2. Estimate ratio of length to width of machine and *block out* on sheet *proportional spaces* for above views.

3. Sketch in all *outlines* **(working on all views at the same time).** Do not attempt to finish one view entirely before working on the other; when a line is placed on one view, place its projection on the other view so that all views are finished at approximately the same time.

4. Sketch in *dimensions, auxiliary,* and *section lines.* The reason for placing on dimension lines while making up the views is, that each detail of the piece as it is drawn may suggest a necessary dimension that perhaps would be overlooked if left until later. A break should be left in each dimension line. No attempt need be made here to distinguish between outline and other lines.

5. Go over the sketch carefully and increase the *weight* of *outlines* so that the construction shows easily.

6. Obtain from the machine with calipers and rule all dimensions already indicated on sketch. **Always** place on *overall* dimensions as a check.

7. Be *liberal* with notes.

(104) **Short cuts.** To save time, the following short cuts are permissible:

1. In drawing objects of familiar shape, wheels, etc., the hub, two spokes, and a short portion of the rim is sufficient.

2. Where objects are symmetrical with respect to a center line, e. g., gate valves, etc., it is sufficient to show only one-half of object, limiting the portion drawn by the center line. The other half may be drawn in when time permits, if desired.

3. Where objects are symmetrical about two center lines at right angles to each other, it will be sufficient to show only one-fourth of the object.

4. Where any part cannot be shown well in detail, e. g., bolts, holes, fasteners, etc., explanatory notes may be substituted—e. g., *⅜″ drill; ¾ x 10 pi. tap; ¾″ Hex. Hd. Mach. Sc.;* etc.

LESSON 12

PERSPECTIVE

(105) Perspective. Tho it is impossible to give here any complete explanation of the principles of perspective, it has been deemed advisable to attempt sufficient explanation to enable engineers, who have no other such opportunity while in college, to understand a few of the basic principles.

It is readily seen that no one view of a working drawing of any object can present to the eye the natural appearance possessed by a crayon or charcoal drawing. The reason is, that in making a working drawing the eye was imagined at an *infinite* distance from the object, an assumption so unnatural as to give rise immediately to results of an unnatural appearance.

(106) Perspective drawing defined. A **perspective drawing** of an object is *such a representation of that object on a given plane or sheet of paper as will present the same appearance as the object itself when the eye is in a certain position with respect to the object.*

The plane on which the perspective drawing is made is called the **picture plane,** and, for reasons which need not be given here, is usually taken *vertically.*

(107) Principles of construction. The principle on which perspective construction is based is as follows: The *vertical picture plane* is placed *between* the *eye* and the *object* (that the drawing may be *smaller* than the object), and **lines of sight** or **visual rays** drawn from the eye to the various points of the object. The *points* in which these lines *pierce* the *picture plane* are respectively the *perspectives* of the corresponding points of the object. If lines be drawn connecting these *piercing points* in their proper order, a perspective drawing of the whole object is obtained.

(108) **Picture plane and position of object.** Since perspective drawings are made mostly from working drawings, the *vertical plane* of *orthographic projection* is used as the *picture plane* and the object placed in the *third angle.*

(109) **Position of point of sight.** The *point* of *sight* is, of course, in *front* of the vertical plane, and may be in either the *first* or *fourth* angles, according to the nature of the view desired; i. e., if it is desired to make a drawing showing the appearance of the object when directly in *front* of it, the point of sight would be in the *fourth* angle.

(110) **Principal point in perspective.** The projection of the *point of sight* on the vertical plane is called the **principal point** in perspective, and is of prime importance in construction. Inasmuch as the vertical projections of points are designated thus, a', b', c', etc., the vertical projection of the *point* of *sight,* S, will be indicated by s'.

(111) **Principal point the vanishing point of lines perpendicular to the picture plane.** It is a familiar fact that as one stands near a long straight section of railroad track the two lines of rails appear to meet off in the distance. So it is with any set of parallel lines; if the eye follows them for a distance—and, when speaking geometrically, we give this distance a value of *infinity*—they all appear to meet in one point. This point we call

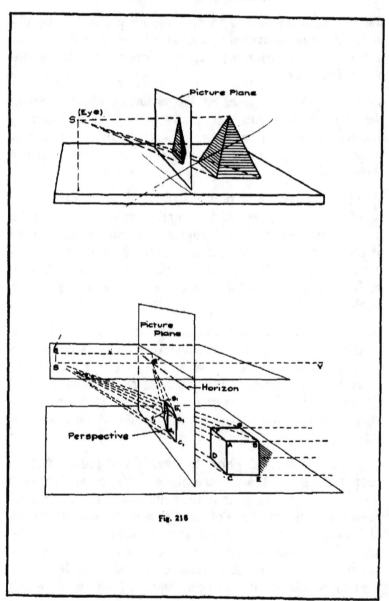

Fig. 216

their **vanishing point.** When our line of sight follows out
these parallel lines to *infinity,* where they appear to meet,
for all practical purposes the *line* of *sight* is *parallel* to
the given set of lines. Reference to Fig. 218 may serve
to make this explanation clearer. Point S represents
the position of the *eye.* A cube AB rests on a horizontal
plane on the other side of the picture plane. The four
parallel edges, AB, CE, etc., of the cube are produced as
indicated by dotted lines to the right; if they are produced an infinite distance they will appear to meet, and
the line of sight from S to the apparent meeting or
vanishing point is the line thru S and V. Then, as explained above, if SV meets AB, CE, etc., at infinity, it
is *parallel* to them. But AB, CE, etc., are *perpendicular*
to the *picture plane;* therefore the line thru S and V out
to this vanishing point is also *perpendicular* to the *picture plane* and must pass thru s', the projection of S on
the picture plane. As viewed from point S, the four
edges, AB, CE, etc., which we have produced to infinity,
do not in reality appear to be parallel lines forming the
edges of a long prism, but seem to represent the four
edges of a long pyramid. To return to the perspective,
suppose we wish to represent this long pyramid on the
picture plane as seen from S. According to Art. 107,
lines are drawn from S to the several points of the
pyramid; the line from S to the imaginary apex at infinity pierces the picture plane at s ; and the lines from
S to A, C, D, and F pierce the picture plane at $a_1 c_1 d_1 f_1$;
then $a_1 c_1 d_1 f_1$ -s' is the perspective of the pyramid. From
this explanation it is seen that the *perspective* of all *lines
perpendicular* to the *picture plane* meet at s', the vertical
projection of the point of sight. For this reason s' is
called the **vanishing point** of perpendiculars. The fact
that perpendiculars do converge at s' affords an easy

method of constructing the perspective of any object when placed in a certain position. Lines from S to the other points of the cube are seen to pierce the picture plane in points on the perspectives of these perpendiculars, giving the figure b_1 a_1 c_1 d_1 f_1 g_1. This figure represents the cube as seen from S. Face ABEC is not visible from S.

(112) **The horizon in perspective.** Any line which is perpendicular to a vertical plane is horizontal. In Fig. 218 the lines AB, CE, etc., are horizontal lines and, when produced an infinite distance, appear to meet in a point on what we commonly call the **horizon.** Then the *line* of *sight* from S to this meeting point becomes a *horizontal line,* and the *perspective* of the *horizon* will be the *horizontal line* drawn thru the point *s'.* The horizontal line lying in the picture plane and passing thru the vertical projections of the point of sight, *s',* is also called the **horizon.**

(113) **One face of the cube coincides with picture plane.** If the cube in Fig. 218 be moved until face ACDF coincides with the picture plane, then this face becomes its own perspective and each line on this face is shown in its true value; i. e., a circle shows as a true circle, etc. From this it follows that the *perspective* of any *circle* whose *plane* is *parallel* to the *picture plane* will be a *true circle;* its diameter will be less or greater than the true diameter, however.

(114) **Mechanical Construction of a Perspective.** Inasmuch as all lines in perspective are shorter than the lines which they represent, except in the case of lines which lie in the picture plane, it will be best to put one face of the object or one face of a circumscribed paral-

lelopiped into coincidence with the picture plane, in order that we may have a foundation of actual measurements on which to base our construction.

(115) **Coordinates.** The position of the point of sight with respect to some chosen point, A, of the object will hereafter be given as follows:

$x =$ distance of point of sight to *right* or *left* of A as x is $+$ or $-$.

$y =$ distance of point of sight *above* or *below* A as y is $+$ or $-$.

$z =$ distance of point of sight *before* the picture plane; e. g., $x = 3''$, $y = -4''$, $z = 6''$ *locates* S $3''$ to the *right* and $4''$ *below* A and $6''$ *before* the picture plane.

Problem 5. **To draw the perspective of a cube $1\frac{1}{4}''$ on edge, one face of the cube coinciding with the picture plane;** $x = -2''$, $y = 1''$, $z = 4''$. A is taken at corner F.

Construction. See Fig. 219.

Draw well toward the top of the sheet a horizontal line G. L., the intersection of the horizontal and vertical planes. Construct in a convenient position the top view *b-ac-df-gj* of the cube; the edge *af*, which represents the top and front edge, coinciding with G. L. $2''$ to the left of point *f* of the *top view* draw a perpendicular to G. L, and $4''$ below G. L., on this perpendicular place the point *s*. The top view, G. L., and *s*, now represent respectively the *cube*, the *picture plane*, and *point of sight*, as they appear looking *down* from above. On the perpendicular from s to G. L. assume point *s'* at any convenient distance   e. g., $\frac{1}{2}''$ below G. L. Then $1''$ below *s'* and limited on the right by *ss'*, construct a *left side view* of the cube and measure from *s'* to the *right* along a horizontal line thru *s'* a distance of $4''$ for point s_1. The *left*

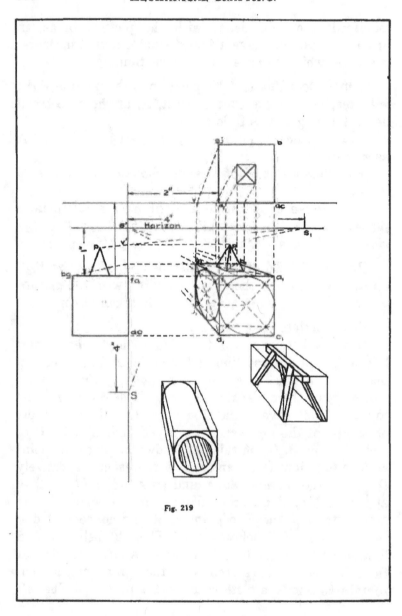

Fig. 219

side, ss′ and *s₁* represent respectively the *cube, picture plane,* and *point of sight* as they appear from the *left.* Then from the top and side views construct the front view $f_1\ a_1\ c_1\ d_1$ of the cube and draw lines from $f_1\ a_1\ c_1$ and d_1 to s'. This figure $s'-f_1\ a_1\ c_1\ d_1$ then is the *perspective* of the *long pyramid* spoken of in Art. 103. Connect *s* with the point *g* of the top view and from the point *v* in which this line intersects G. L. drop a perpendicular to G. L., until it intersects the two lines $f_1\ s'$ and $d_1\ s'$; $g_1\ j_1$ is then the perspective of the edge GJ of the cube. A horizontal line from g_1 produced until it intersects $a_1\ s'$ at b_1 completes the perspective of the cube. It is easily understood that the line *sg* represents a *line* of *sight* from S to G and was drawn to ascertain the point V at which this line of sight pierces the picture plane, or rather to determine the distance *vd* to the left of the edge FD of the cube at which that line of sight pierces the picture plane.

It is desired to place a small pyramid on top of the cube, the edges of its base parallel to the edges of the cube and its axis passing thru the center of the top face. Construct the top and left side views in place and proceed with the construction of the base as shown. After the perspective of the base is drawn the two diagonals may be drawn to determine the position of the axis. Connect s_1 and *p;* $s_1\ p$ intersects *ss′* at v'; a horizontal line thru v' intersects the axis of the pyramid at p_1, and the pyramid may be completed. The *left side* view and *s,* enable us to determine the distance *above* any line of the face FACD at which lines of sight pierce the picture plane, $v'\ a$ being the distance above FA at which *SP* pierces this plane.

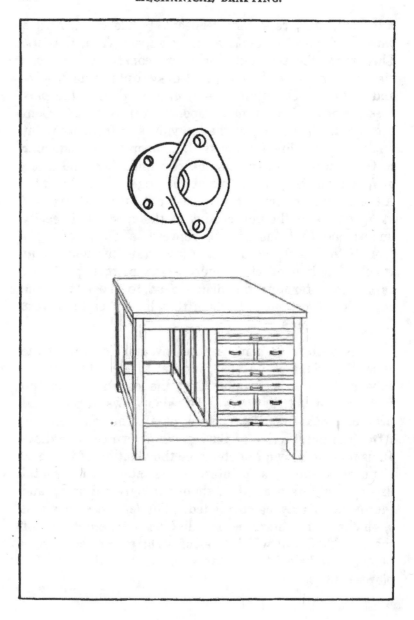

(116) **Circles in perspective.** The **8 point method** may be used in constructing circles in perspective. This is illustrated on the cube in Fig. 219. The diagonals can easily be drawn and the points of tangency of the perspective with the upper and lower lines f_1 g_1 and d_1 j_1 determined by drawing a vertical line thru the intersection of the diagonals.

(117) **Irregularly shaped objects.** Any irregularly shaped object may be easily drawn in perspective by first enclosing the object in a parallelopiped and referring the several constructions of object to lines of the parallelopiped.

(118) **Position of point of sight.** Considerable care and judgment must be used in placing the point of sight, for it is easily understood that a house viewed from a point only two feet in front of it would look absurd; however, its perspective can be constructed as easily under such circumstances as any other. It is well to estimate approximately from what particular position we would likely view that object to obtain the best view; taking into account the size of the object in this estimate. The point of sight may then be placed accordingly. For large objects a safe rule is to place the point of sight in front of the object a distance equal to twice the greatest dimension. For smaller objects we may increase this to 4 or 5 times the greatest dimension.

THE ELLIPSE

(119) **Definition:** *An ellipse is a curve generated by the motion of a point which moves so that the sum of its distances from two fixed points is constant.* For example, in Fig. 220, the sum, x + y, of the distances

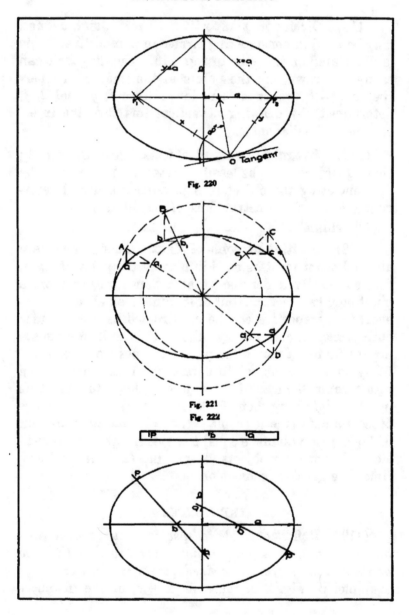

Fig. 220

Fig. 221
Fig. 222

from any point O on the ellipse to the two fixed points, F_1 and F_2 is constant and equal to 2a.

Besides being considered a mathematical curve generated according to a certain law, the ellipse may be considered the curve which is cut from the surface of a right circular cone by a plane which intersects all of the elements and is oblique to the axis. Also as the orthographic projection of a circle which is oblique to the plane of projection.

The *long diameter* of the ellipse is known as the **major axis** and the *short diameter* as the **minor axis**; in analytic geometry these axes are given values of 2a and 2b, Fig. 220.

Construction. The ellipse figures so prominently in drafting that it will be well to give several methods of construction, both exact and approximate.

Exact Method. (1) From the law according to which the curve is generated it has been found possible to construct the ellipse accurately as follows. With point O, Fig. 221, as a center and the axes as diameters, describe two circles. Then from O draw any number of radii of the large circle, e. g., OA, OB, OC, OD, etc. The vertices a, b, c, d, etc., of the right angles of the right triangles, Aaa_1, Bbb_1, etc., are points of the required ellipse and the curve may be traced thru these points either freehand or by means of a universal curve.

Exact method (2) Trammel method. If from any point P, Fig. 222, on the edge of a strip of paper or ruler the *semi minor* and *semi major axes* be measured to points b and a, and this strip or ruler placed over the axes and moved so that point b is always on the major axis and a on the minor axis, the successive positions of

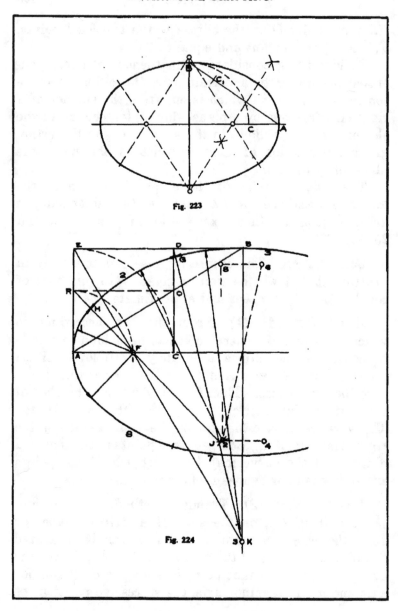

Fig. 223

Fig. 224

point P are points of the required ellipse. These positions of point P may be marked with a pencil or needle point and the ellipse traced thru them.

Approximate method (1)—4 center method. Connect point B, Fig. 223, with point A. Then with O as a center and OB as a radius describe the arc cutting OA at C; lay off from B, on BA, the distance CA, to C_1. The perpendicular bisector of C_1A locates two of the desired centers and the curve may be drawn with the compass as shown.

Approximate method (2)—8 center method. Construction. Connect points A and B and draw the lines AE and BE, Fig. 224. Then describe the quadrant EC and erect the perpendicular CD. From point O in which CD intersects AB draw OR. The arc RF locates center No. 1. EF produced locates center No. 3 at K. Connect D and K and produce RF to J; with G and H as centers and GJ as a radius describe arcs intersecting at center No. 2. Centers, 4, 6, and 8 may then be located from 2. It will be noted that center No. 2 does not lie on the radius GK; however, it is so small a distance from GK that no irregularity can be detected in the curves at G.

LESSON 13

PHOTOGRAPHIC REPRODUCTION

(120) **Blueprinting.** Blueprinting is in short a process of simple photographic reproduction on sensitized paper, of drawings which have been made on some translucent material; this material may be *tracing cloth, tracing paper,* or ordinary paper *oiled* after the drawing has been finished. In common practice the process is somewhat rough as one would infer from a glance at the average print; however, with care it can be carried nearly to the same limits of refinement as other photographic printing.

BLUEPRINT PAPERS

(121) Occasions may arise when it is necessary to sensitize paper for blueprinting; however, unless absolutely necessary no draftsman should ever bother to coat his own paper, for it is a tedious and most unsatisfactory process. The machine coated papers sold by any of the instrument companies in 10 or 50 yard rolls of any width and of any desired thickness or quality of paper is cheap, keeps well for months in a tin tube, and always gives better results than paper coated by the amateur. For prints which must stand extra hard use, either mounted paper (cloth backed paper) or blueprint cloth (a smooth hard surfaced sensitized cloth) should be used; the latter, however, seldom gives the sharp detail obtainable on paper. Below is tabulated information that will be of value in ordering papers or cloth.

Blue Print Paper and Cloth

PAPER		CLOTH	
WEIGHT	USE	WEIGHT	USE
Extra thin	For sending thru mail. Not satisfactory for shop use, too thin.	Extra thin	Large prints which are to receive extra hard wear.
Thin	For large prints that would be too bulky on heavier paper.	Medium	Small maps, moderate sized prints that receive extra wear.
Medium thick	Best for all ordinary use, shop, construction work, etc.		
Thick	For durable small prints; maps, land plots, etc.		

Printing Speed in Bright Sunlight

Regular	Rapid	Extra Rapid	Elec. Rapid
4 min.	2 min.	40 sec.	25 sec.

In ordering paper be sure to state the printing speed desired; e. g., 1 roll; 50 yds. x 36 in., Extra Thin, Electric Rapid, Blue Print Paper.

TO SENSITIZE PAPER

(122) **Paper.** Only "unsized" and well washed papers are suitable for blueprinting. The *size* used on many papers to give it a glossy and easy writing surface discolors the blueprint solution immediately. Likewise paper from which the sulphur, used in its manufacture, has not been well washed will discolor. Practically any unsized "bond" or "parchment" paper will be found satisfactory for printing.

Solution. The formula in most common use for the sensitizing solution is: (1) Red prussiate of potash, 1 oz.; water (distilled), 4 oz.; (2) double citrate of iron and ammonia, 1 oz.; water (distilled), 4 oz. As long as these solutions are kept separate sunlight has no effect upon them. However, the second solution should be kept in a well stoppered bottle of dark colored glass.

To sensitize the paper, mix *equal* volumes of the above solutions and apply either with a camel's hair brush, brushing first horizontally, then vertically, to insure even coating, or float the paper for a few seconds in a shallow granite pan partially filled with the solution, and hang by one corner to dry. This sensitizing must of course be done in a dark room. If the solution of citrate of iron is kept too long it may mold and spoil; hence, as the crystals dissolve quite readily it may be best to make this solution only when it is to be used and then only what is needed. The bottle in which the citrate is kept should be glass stoppered to prevent moisture from melting down the crystals. The prussiate does not dissolve so readily and as it does not spoil it can be kept in solution in any quantity.

VANDYKE SOLAR PAPER

(123) Vandyke Solar Paper, sometimes called "Brown Print" paper, is a brown paper used in making negatives for *positive printing*. A print is made on Vandyke paper from a tracing, the *inked side* of the tracing being in this case turned to the paper, so that a reversed print is obtained. The lines of the drawing show up *white* on a deep *brown* background. This paper is then rubbed with oil to make it more transparent and positive blueprints are made from it, the *brown side* of the negative being turned **toward** the blueprint paper. In this final print the lines of the drawing show up *blue* on a *white* background instead of the reverse in direct printing from the tracing.

WASHING AND FIXING VANDYKES

(124) Vandyke paper has been sufficiently exposed when the surface not covered by the black lines of the tracing has turned a *light bronze* color. After washing for a few minutes, **face down** in the tank, the print should be **fixed** with a solution of 1 oz. of hyposulphite of soda to 1 qt. of water. The print may be laid on a board and the solution applied with a brush or better still the print may be floated **face down**, in a granite pan partially filled with the hypo solution; one brushing or a few seconds floating is sufficient. The print should then be washed again **face down**, to remove the surplus hypo.

TO TRANSPARENTIZE VANDYKES

(125) It is possible to obtain positive prints from unoiled Vandykes; however, if the negative has been rendered more transparent by oiling the printing time will be materially reduced. Any clear oil or white grease will answer for this purpose, *white tube vaseline* being perhaps the most convenient. The following formula gives a transparentizing oil that works well: 4 oz. banana oil, 10c tube white vaseline. (Mix the two by heating slightly and keep in a stoppered bottle.) The banana oil furnishes a quick "drier" and the vaseline a permanent oil. If there is much transparentizing to be done it is well to keep a ball of waste, soaked in the above solution, in a covered tin can. Never apply more grease than will dry in a few minutes and be sure to oil the side of the paper which is to be turned **toward** the light. Unless necessary **never** use **paraffin** for transparentizing; it renders the paper very brittle and any wrinkles in a paraffined negative show plainly on the print.

POSITIVES ON OLD VANDYKE PAPER

(126)   In making positive prints on old Vandyke paper some little care must be exercised. When printed and washed the unexposed or white parts turn decidedly yellow; this can be prevented if the print is merely dipped in the water to wet the surface and start the printing out and immediately floated on the fixer; the fixer will print out the lines and prevent the background from turning yellow. These precautions are not necessary in making negatives on old paper, for tho unexposed parts may turn yellow the negative will print well when oiled.

BLUEPRINTING FROM TYPEWRITTEN SHEETS

(127)   To obtain clear sharp prints from typewritten sheets a moderately thin hard surfaced paper and new black typewriter ribbon should be used. If there is much work to be done it will be well to obtain an extra heavily inked ribbon. In typewriting place *under* the the paper a sheet of black carbon paper with carbon face **toward** the paper; thus an impression is obtained on both sides of the paper. Use each sheet of carbon paper only **once** for this purpose.

In oiling, one may not rub these sheets, as the carbon will smear. Instead, lay over the typewritten sheet a square of oily cotton flannel, then a sheet of heavy paper and roll with a small picture mounting roll. Or better, if time permits, lay pieces of oily cotton cloth between the sheets and weight down with a heavy book for several hours. Arrange sheets and cloth as follows: Paper, cloth, two sheets of paper, cloth, two sheets of paper, cloth, etc. In printing from these sheets, over expose the paper slightly and wash in water to which hydrogen peroxide has been added in the proportion of ½ tea-spoonful to 2 gallons water; the hydrogen peroxide will bring out the over exposed parts and deepen the blue. The above solution can be used to advantage in washing any blueprints; a sharper contrast between the white lines and blue background can be obtained.

PRINTING FROM OLD BLUEPRINTS

(128) Occasionally reproductions of drawings are wanted when the tracings are not available. By use of the "Direct Copier" of the Frederick Post Co., Chicago, an old blueprint may be rendered sufficiently dense to act as a good negative. This "Direct Copier" consists of two concentrating solutions to be applied to the old blueprint to deepen the blue; the transparentizing oil must then be used to clear up the white lines. If the "Direct Copier" is not available and there is not sufficient time for a tracing a fairly good positive may be made from a blueprint by merely transparentizing it.

PRINTING FROM HEAVY CARDBOARD

(129) If it is desired to make a blueprint of a drawing mounted or printed on heavy cardboard or of a drawing on mounted paper, the face of the drawing should be soaked with alcohol and immediately clamped in the printing-frame. The alcohol will not evaporate while closed in the frame nor will it dissolve the blueprint solution if it should soak thru the cardboard. The length of time required for printing will have to be learned from experiment.

PRINTING FROM COORDINATE PAPER

(130) Coordinate paper printed in **red** gives better blueprints than the paper printed in blue or green. Always transparentize the coordinate paper before printing; it will save time in printing and give better results.

(131) **Positive blueprinting of typewritten sheets.** Excellent negatives for positive printing of typewritten sheets may be made from *new black carbon paper* as follows: Remove the ribbon from the typewriter and place the carbon paper in the machine, **face up,** with a sheet of thin hard surfaced paper over it. A reversed impression will of course be obtained on the back of this sheet of paper and the better the quality of paper the more clear cut will be the letters on the carbon sheet. The reason for placing the carbon with face **to** the cover sheet is to obtain such an impression as to make it possible to place the carbon side of the paper **toward** the *glass* of the printing frame instead of toward the *blueprint paper*.

Handle the carbon paper very carefully; a finger mark or smudge may easily ruin the negative.

The time for printing will have to be determined by experiment; it should be somewhat longer than for printing from tracings.

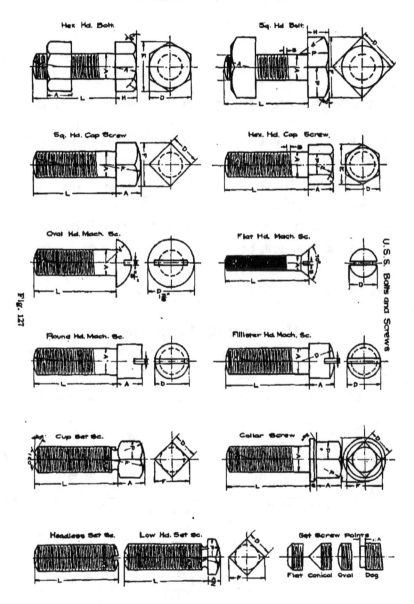

Fig. 127

U.S.S. Bolts and Screws

REFERENCE TABLES

KEY FOR THE FOLLOWING TABLES OF BOLTS, NUTS, ETC.

A=Outside Diameter of Threads and Thickness of Nut.

B=Threads per Inch.

C=Tapping Drill.

D=Across Flats.

E=Across Corners, Hex.

F=Across Corners, Sq.

G=Thickness of Collar.

H=Thickness of Head.

I=Across Flats.

J=Thickness Head and Nut.

K=Diameter Collar.

S=Width of Slot.—Decimals.

X=Angle of Head.

Hexagonal-Head Cap-Screw

A	1/4	5/1·	3/8	7/16	1/2	9/16	5/8	3/4	7/8
B	20	18	16	14	12	12	11	10	9
D	7/16	1/2	9/16	5/8	3/4	13/16	7/8	1	1- 1/8
E	1/2	37/64	41/64	23/32	55/64	15/16	33/64	1-5/32	1-19/64

Square-Head Cap-Screw

A	1/4	5/16	3/8	7/16	1/2	9/16	5/8	3/4	7·8
B	2·	18	16	14	12	12	11	10	9
D	3/8	7/16	1/2	9/16	3/8	11/16	3/4	7/8	1- 1/8
F	17/32	39/64	45/64	51/64	7/8	31·32	1- 1/16	1-15/64	1-19/32

Iron Set-Screw

A & D	1/4	5·16	3/8	7/16	1/2	9/16	5/8	3/4	7·8
B	20	18	16	14	12	12	11	10	9

U. S. Standard Bolts and Nuts

			Rough				Finished	
A	B	C	D	E	F	H	I	J
1/4	20	10	1/2	37/64	7/10	1/4	7/16	3/16
5/16	18		1/4	19/32	11/16	10/12	19/64	17/32	1/4
3/8	16	19/64	11/16	51/n4	63/64	11/32	5/8	5/16
7/16	14	23/64	25/32	9/10	1- 7/64	25/64	23/32	3/8
1/2	13	13/32	7/8	1	1-15/64	7/16	13/16	7/16
9/16	12	15/32	31/32	1- 1/8	1-23/64	31/64	29/32	1/2
5/8	11	33/64	1- 1/16	1- 7/32	1- 1/2	17/32	1	9/16
3/4	10	5/8	1- 1/4	1- 7/16	1-49/64	5/8	1- 3/16	11/16
7/8	9	47/64	1- 7/16	1-21/32	2- 1/32	23/32	1- 3/8	13/16
1	8	27/32	1- 5/8	1- 7/8	2-19/64	13/16	1- 9/16	15/16
1-1/8	7	61/64	1-13/16	2- 3/32	2- 9/16	29/32	1- 3/4	1- 1/16
1-1/4	7	1- 5/64	2	2- 5/16	2-53/64	1	1-15/16	1- 3/16
1-3/8	6	1-11/64	2- 3/16	2-17/32	3- 3/32	1- 3/32	2- 1/8	1- 5/16
1-1/2	6	1-19/64	2- 3/8	2- 3/4	3-23/64	1- 3/16	2- 5/16	1- 7/16
1-5/8	5-1/2	1-25/64	2- 9/16	2-31/32	3- 5/8	1- 9/32	2- 1/2	1- 9/16
1-3/4	5	1- 1/2	2- 3/4	3- 3/16	3-57/64	1- 3/8	2-11/16	1-11/16
1-7/8	5	1- 5/8	2-15/16	3-13/32	4- 5/32	1-15/32	2- 7/8	1-13/16
2	4-1/2	1-23/32	3- 1/8	3-19/32	4-27/64	1- 9/16	3- 1/16	1-15/16
2-1/4	4-1/2	1-31/32	3- 1/2	4- 1/32	4-61/64	1- 3/4	3- 7/16	2- 3/16
2-1/2	4	2- 3/16	3- 7/8	4-15/32	5-31/64	1-15/16	3-13/16	2- 7/16
2-3/4	4	2- 7/16	4- 1/4	4-29/32	6	2- 1/8	4- 3/16	2-11/16
3	3-1/2	2-41/64	4- 5/8	5-11/32	6-17/32	2- 5/16	4- 9/16	2-15/16
3-1/4	3-1/2	2-57/64	5	5-25/32	7- 1/16	2- 1/2	4-15/16	3- 3/16
3-1/2	3-1/4	3- 1/8	5- 3/8	6-13/64	7-39/64	2-11/16	5- 5/16	3- 7/16
3-3/4	3	3-21/64	5- 3/4	6- 5/8	8- 1/8	2- 7/8	5-11/16	3-11/16
4	3	3-37/64	6- 1/8	7- 1/16	8-41/64	3- 1/16	6- 1/16	3-15/16
4-1/4	2-7/8	3-13/16	6- 1/2	7- 1/2	9- 3/16	3- 1/4	6- 7/16	4- 3/16
4-1/2	2-3/4	4- 3/64	6- 7/8	7-15/16	9- 3/4	3- 7/16	6-13/16	4- 7/16
4-3/4	2-5/8	4- 9/32	7- 1/4	8- 3/8	10- 1/4	3- 5/8	7- 3/16	4-11/16
5	2-1/2	4- 1/2	7- 5/8	8-13/16	10-49/64	3-13/16	7- 9/16	4-15/16
5-1/4	2-1/2	4- 3/4	8	9-15/64	11- 5/16	4	7-15/16	5- 3/16
5-1/2	2-3/8	4-63/64	8- 3/8	9-11/16	11-27/32	4- 3/16	8- 5/16	5- 7/16
5-3/4	2-3/8	5-15/64	8- 3/4	10- 3/32	12- 3/8	4- 3/8	8-11/16	5-11/16
6	2-1/4	2-29/64	9- 1/8	10-17/32	12-15/16	4- 9/16	9- 1/16	5-15/16

Flat-Head Machine Screw

A	1/8	3/16	1/4	5/16	3/8	7/16	1/2	9/16	5/8	3/4
B	40	24	2()	18	16	14	12	12	11	10
D	1/4	3/8	15/32	5/8	3/4	13/16	7/8	1	1-1/8	1-3/8
S	.028	.035	.051	.057	.064	.081	.081	.091	.114	.128
X	70	70	70	70	70	70	70	70	70	70

Round and Fillister Head Machine Screw

A	1/8	3/16	1/4	5/16	3/8	7/16	1/2	9/16	5/8	3/4
B	40	24	20	18	16	14	12	12	11	10
D	3/16	1/4	3/8	7/16	9/16	5/8	3/4	13/16	7/8	1
S	.028	.035	.051	.057	.064	.081	.081	.091	.114	.128

Collar-Screw

A	1/8	3/16	1/4	5/16	3/8	7/16	1/2	9/16	5/8	3/4
B	40	24	20	18	16	14	12	12	11	10
D	1/8	3/16	1/4	5/16	3/8	7/16	1/2	9/16	5/8	3/4
F	11/64	17/64	11/32	7/16	17/32	39/64	11/16	51/64	7/8	1-1/16
C	1/4	11/32	7/16	1/2	5/8	11/16	15/16	15/16	1	1-1/4
G	1/32	3/64	1/16	5/64	3/32	7/64	1/8	9/64	5/32	3/16

Pipe Threads

Diam.	Thds. Per In.	Diam. Drill	Diam.	Thds. Per In.	Diam. Drill
1/8	27	21/64	1-1/4	11-1/2	1-15/38
1/4	18	29/64	1-1/2	11-1/2	1-23/32
3/8	18	19/32	2	11-1/2	2- 3/16
1/2	14	25/38	2-1/2	8	2-11/16
3/4	14	15/16	3	8	3- 5/16
1	11-1/2	1-3/16	3-1/2	8	3-13/16

Standard taper of pipe threads is, 1 inch in 16, or 3/4 inch to 1 foot.

U. S. STANDARD SCREW THREADS.

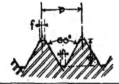

FORMULA

$$p = pitch = \frac{1}{No, threads\ per\ inch}$$

$$d = depth = p \times .6495$$

$$f = flat = \frac{p}{8}$$

Diameter of Screw.	Threads per Inch.	Diam. at Root of Thread.	Width of Flat.
¼	20	.185	.0063
⁵⁄₁₆	18	.2403	.0069
⅜	16	.2936	.0078
⁷⁄₁₆	14	.3447	.0089
½	13	.4001	.0096
⁹⁄₁₆	12	.4542	.0104
⅝	11	.5069	.0114
¾	10	.6201	.0125
⅞	9	.7307	.0139
1	8	.8376	.0156
1⅛	7	.9394	.0179
1¼	7	1.0644	.0179
1⅜	6	1.1585	.0208
1½	6	1.2835	.0208
1⅝	5½	1.3888	.0227
1¾	5	1.4902	.0250
1⅞	5	1.6152	.0250
2	4½	1.7113	.0278
2¼	4½	1.9613	.0278
2½	4	2.1752	.0313
2¾	4	2.4252	.0313
3	3½	2.6288	.0357
3¼	3½	2.8788	.0357
3½	3¼	3.1003	.0385
3¾	3	3.3170	.0417
4	3	3.5670	.0417
4¼	2⅞	3.7982	.0435
4½	2¾	4.0276	.0455
4¾	2⅝	4.2551	.0476
5	2½	4.4804	.0500
5¼	2½	4.7304	.0500
5½	2⅜	4.9530	.0526
5¾	2⅜	5.2030	.0526
6	2¼	5.4226	.0556

FOR TAPS WITH U. S. STANDARD THREADS.

Size of Tap.	No. of Thds.	Size of Drill.	Size of Tap.	No. of Thds.	Size of Drill.	Size of Tap.	No. of Thds	Size of Drill.	Size of Tap.	No. of Thds.	Size of Drill.
$\frac{1}{4}$	20	$\frac{7}{16}$ in.	$\frac{11}{16}$	11	$\frac{33}{64}$	$1\frac{1}{4}$	7	$1\frac{5}{64}$	$2\frac{1}{8}$	$4\frac{1}{2}$	$1\frac{27}{64}$
$\frac{5}{16}$	18	C	$\frac{3}{4}$	10	$\frac{5}{8}$	$1\frac{3}{8}$	6	$1\frac{11}{64}$	$2\frac{1}{4}$	$4\frac{1}{2}$	$1\frac{31}{64}$
$\frac{3}{8}$	16	N	$\frac{13}{16}$	10	$\frac{11}{16}$	$1\frac{1}{2}$	6	$1\frac{11}{64}$	$2\frac{3}{8}$	4	$2\frac{1}{16}$
$\frac{7}{16}$	14	S	$\frac{7}{8}$	9	$\frac{47}{64}$	$1\frac{5}{8}$	$5\frac{1}{2}$	$1\frac{29}{64}$	$2\frac{1}{2}$	4	$2\frac{3}{16}$
$\frac{1}{2}$	13	$\frac{27}{64}$ in.	$\frac{15}{16}$	9	$\frac{51}{64}$	$1\frac{3}{4}$	5	$1\frac{1}{2}$			
$\frac{9}{16}$	12	$\frac{31}{64}$ in.	1	8	$\frac{53}{64}$	$1\frac{7}{8}$	5	$1\frac{5}{8}$			
$\frac{5}{8}$	11	$\frac{33}{64}$ in.	$1\frac{1}{8}$	7	$\frac{61}{64}$	2	$4\frac{1}{2}$	$1\frac{11}{64}$			

Laying Out Angles with a 2-ft. Rule

Open a 2-ft rule until the open ends are as far apart as the distance shown in table below for the desired angle. To measure an angle reverse the operation.

Degrees	Inches	Degrees	Inches	Degrees	Inches
1	.21	15	3.12	55	11.08
2	.422	20	4.17	60	12.0
3	.633	25	5.21	65	12.89
4	.837	30	6.21	70	13.76
5	1.04	35	7.20	75	14.61
7.5	1.57	40	8.21	80	15.43
10	2.09	45	9.20	85	16.21
14.5	3.015	50	10.12	90	16.97

DECIMAL EQUIVALENTS

Of 8ths, 16ths, 32nds and 64ths of an inch

8ths.

$\frac{1}{8}$ = .125
$\frac{1}{4}$ = .250
$\frac{3}{8}$ = .375
$\frac{1}{2}$ = .500
$\frac{5}{8}$ = .625
$\frac{3}{4}$ = .750
$\frac{7}{8}$ = .875

16ths.

$\frac{1}{16}$ = .0625
$\frac{3}{16}$ = .1875
$\frac{5}{16}$ = .3125
$\frac{7}{16}$ = .4375
$\frac{9}{16}$ = .5625
$\frac{11}{16}$ = .6875
$\frac{13}{16}$ = .8125
$\frac{15}{16}$ = .9375

32ds.

$\frac{1}{32}$ = .03125
$\frac{3}{32}$ = .09375

$\frac{5}{32}$ = .15625
$\frac{7}{32}$ = .21875
$\frac{9}{32}$ = .28125
$\frac{11}{32}$ = .34375
$\frac{13}{32}$ = .40625
$\frac{15}{32}$ = .46875
$\frac{17}{32}$ = .53125
$\frac{19}{32}$ = .59375
$\frac{21}{32}$ = .65625
$\frac{23}{32}$ = .71875
$\frac{25}{32}$ = .78125
$\frac{27}{32}$ = .84375
$\frac{29}{32}$ = .90625
$\frac{31}{32}$ = .96875

64ths.

$\frac{1}{64}$ = .015625
$\frac{3}{64}$ = .046875
$\frac{5}{64}$ = .078125
$\frac{7}{64}$ = .109375
$\frac{9}{64}$ = .140625
$\frac{11}{64}$ = .171875
$\frac{13}{64}$ = .203125

$\frac{15}{64}$ = .234375
$\frac{17}{64}$ = .265625
$\frac{19}{64}$ = .296875
$\frac{21}{64}$ = .328125
$\frac{23}{64}$ = .359375
$\frac{25}{64}$ = .390625
$\frac{27}{64}$ = .421875
$\frac{29}{64}$ = .453125
$\frac{31}{64}$ = .484375
$\frac{33}{64}$ = .515625
$\frac{35}{64}$ = .546875
$\frac{37}{64}$ = .578125
$\frac{39}{64}$ = .609375
$\frac{41}{64}$ = .640625
$\frac{43}{64}$ = .671875
$\frac{45}{64}$ = .703125
$\frac{47}{64}$ = .734375
$\frac{49}{64}$ = .765625
$\frac{51}{64}$ = .796875
$\frac{53}{64}$ = .828125
$\frac{55}{64}$ = .859375
$\frac{57}{64}$ = .890625
$\frac{59}{64}$ = .921875
$\frac{61}{64}$ = .953125
$\frac{63}{64}$ = .984375

WEIGHTS
OF SQUARE AND ROUND BARS OF CARBON STEEL IN POUNDS PER LINEAL FOOT.

Weight of 1 cubic inch = .285 lbs.

The following tables are calculated from the unit of 1 cubic inch = .3 lbs. which in practice has proved very accurate as nearly all steel is finished slightly full to dimensions.

Thickness or Diameter in Inches.	Weight of Square Bar One Foot Long.	Weight of Round Bar One Foot Long.	Thickness or Diameter in Inches.	Weight of Square Bar One Foot Long.	Weight of Round Bar One Foot Long.	Thickness or Diameter in Inches.	Weight of Square Bar One Foot Long.	Weight of Round Bar One Foot Long.
$\frac{1}{16}$	.014	.011	$1\frac{3}{4}$	11.02	8.65	$3\frac{7}{16}$	42.5	33.4
$\frac{1}{8}$	.056	.044	$1\frac{13}{16}$	11.82	9.28	$3\frac{1}{2}$	44.1	34.6
$\frac{3}{16}$	.126	.099	$1\frac{7}{8}$	12.65	9.94	$3\frac{9}{16}$	45.7	35.8
$\frac{1}{4}$	.225	.177	$1\frac{15}{16}$	13.51	10.61	$3\frac{5}{8}$	47.3	37.1
$\frac{5}{16}$	.351	.276	2	14.4	11.3	$3\frac{11}{16}$	48.9	38.4
$\frac{3}{8}$	.506	.397	$2\frac{1}{16}$	15.3	12.0	$3\frac{3}{4}$	50.6	39.7
$\frac{7}{16}$	.689	.541	$2\frac{1}{8}$	16.2	12.7	$3\frac{13}{16}$	52.3	41.0
$\frac{1}{2}$	.900	.707	$2\frac{3}{16}$	17.2	13.5	$3\frac{7}{8}$	54.0	42.4
$\frac{9}{16}$	1.13	.895	$2\frac{1}{4}$	18.2	14.3	$3\frac{15}{16}$	55.8	43.8
$\frac{5}{8}$	1.40	1.10	$2\frac{5}{16}$	19.2	15.1	4	57.6	45.2
$\frac{11}{16}$	1.70	1.33	$2\frac{3}{8}$	20.3	15.9	$4\frac{1}{16}$	59.4	46.6
$\frac{3}{4}$	2.02	1.59	$2\frac{7}{16}$	21.4	16.8	$4\frac{1}{8}$	61.2	48.1
$\frac{13}{16}$	2.37	1.86	$2\frac{1}{2}$	22.5	17.6	$4\frac{3}{16}$	63.1	49.5
$\frac{7}{8}$	2.75	2.16	$2\frac{9}{16}$	23.6	18.52	$4\frac{1}{4}$	65.0	51.0
$\frac{15}{16}$	3.16	2.48	$2\frac{5}{8}$	24.8	19.4	$4\frac{5}{16}$	66.9	52.5
1	3.60	2.82	$2\frac{11}{16}$	26.0	20.4	$4\frac{3}{8}$	68.9	54.1
$1\frac{1}{16}$	4.06	3.19	$2\frac{3}{4}$	27.2	21.3	$4\frac{7}{16}$	70.9	55.6
$1\frac{1}{8}$	4.55	3.57	$2\frac{13}{16}$	28.4	22.3	$4\frac{1}{2}$	72.9	57.2
$1\frac{3}{16}$	5.07	3.98	$2\frac{7}{8}$	29.7	23.3	$4\frac{9}{16}$	74.9	58.8
$1\frac{1}{4}$	5.62	4.41	$2\frac{15}{16}$	31.0	24.4	$4\frac{5}{8}$	77.0	60.4
$1\frac{5}{16}$	6.20	4.87	3	32.4	25.4	$4\frac{11}{16}$	79.1	62.0
$1\frac{3}{8}$	6.80	5.34	$3\frac{1}{16}$	33.7	26.5	$4\frac{3}{4}$	81.2	63.7
$1\frac{7}{16}$	7.43	5.84	$3\frac{1}{8}$	35.1	27.6	$4\frac{13}{16}$	83.3	65.4
$1\frac{1}{2}$	8.10	6.36	$3\frac{3}{16}$	36.5	28.7	$4\frac{7}{8}$	85.5	67.1
$1\frac{9}{16}$	8.78	6.90	$3\frac{1}{4}$	38.0	29.8	$4\frac{13}{16}$	87.7	68.8
$1\frac{5}{8}$	9.50	7.46	$3\frac{5}{16}$	39.5	31.0	5	90.0	70.6
$1\frac{11}{16}$	10.24	8.05	$3\frac{3}{8}$	41.0	32.2	$5\frac{1}{16}$	92.3	72.4

THICKNESS AND WEIGHT OF SHEET STEEL AND IRON.

Number of Gage.	Approximate Thickness		Weight Per Sq. Foot.		*Overweight
	Fractions.	Decimals.	Steel.	Iron.	Up to 75 in. Wide.
0000000	½	.5	20.320	20.00	5 per cent.
000000		.46875	19.050	18.75	
00000		.4375	17.780	17.50	6 " "
0000		.40625	16.510	16.25	
000		.375	15.240	15.00	7 " "
00		.34375	13.970	13.75	
0		.3125	12.700	12.50	8 " "
1		.28125	11.430	11.25	
2		.26562	10.795	10.625	Up to 50 in.
3	¼	.25	10.160	10.00	Wide.
4		.23437	9.525	9.375	
5		.21875	8.890	8.75	
6		.20312	8.255	8.125	7 per cent.
7		.1875	7.620	7.5	
8		.17187	6.985	6.875	8½ " "
9		.15625	6.350	6.25	
10		.14062	5.715	5.625	10 " "
11	⅛	.125	5.080	5.00	
12		.10937	4.445	4.375	
13		.09375	3.810	3.75	
14		.07812	3.175	3.125	
15		.07031	2.857	2.812	
16		.0625	2.540	2.50	
17		.05625	2.286	2.25	
18		.05	2.032	2.	
19		.04375	1.778	1.75	
20		.0375	1.524	1.50	
21		.03437	1.397	1.375	
22		.03125	1.270	1.25	
23		.02812	1.143	1.125	
24		.025	1.016	1.	
25		.02187	1.389	.875	
26		.01875	.762	.75	
27		.01718	.698	.687	
28		.01562	.635	.623	
29		.01406	.571	.562	
30		.0125	.508	.5	
31		.01093	.694	.437	
32		.01015	.413	.406	
33		.00937	.381	.375	
34		.00859	.349	.343	
35		.00781	.317	.312	
36		.00703	.285	.281	
37		.00664	.271	.265	
38		.00625	.254	.25	

SIZES OF NUMBERS OF THE U. S. STANDARD GAGE

Number of Gage.	Approximate Thickness in Fractions of an Inch.	Approximate Thickness in Decimal Parts of an Inch.	Weight per Square Foot in Ounces. Avoirdupois.	Weight per Square Foot in Pounds. Avoirdupois.
16	$\frac{1}{16}$	.0625	40	2.5
17	$\frac{18}{320}$	.05625	36	2.25
18	$\frac{1}{20}$	.05	32	2.
19	$\frac{14}{320}$	.04375	28	1.75
20	$\frac{3}{80}$	.0375	24	1.50
21	$\frac{11}{320}$	.034375	22	1.375
22	$\frac{1}{32}$	.03125	20	1.25
23	$\frac{9}{320}$	.028125	18	1.125
24	$\frac{1}{40}$	.025	16	1.
25	$\frac{7}{320}$	.021875	14	.875
26	$\frac{3}{160}$	.01875	12	.75
27	$\frac{11}{640}$	.0171875	11	.6875
28	$\frac{1}{64}$	.015625	10	.625
29	$\frac{9}{640}$	.0140625	9	.5625
30	$\frac{1}{80}$	.0125	8	.5
31	$\frac{7}{640}$	.0109375	7	.4375
32	$\frac{13}{1280}$	.01015625	6½	.40625
33	$\frac{3}{320}$	.009375	6	.375
34	$\frac{11}{1280}$	.00859375	5½	.34375
35	$\frac{5}{640}$	.0078125	5	.3125
36	$\frac{9}{1280}$	.00703125	4½	.28125
37	$\frac{17}{2560}$	.006640625	4¼	.265625
38	$\frac{1}{160}$	.00625	4	.25

DIFFERENT STANDARDS FOR WIRE GAGE.

Number of Wire Gage.	American or Brown & Sharpe.	Birmingham or Stubs' Wire.	Washburn & Moen Mfg. Co. Worcester. Mass.	Imperial Wire Gage.	Stubs' Steel Wire	U. S. Standard for Plate.	Number of Wire Gage.
000000	.....	.....	.....	.464	....	.46875	000000
00000	.....	.....	.....	.432	....	.4375	00000
0000	.46	.454	.3938	.400	....	.40625	0000
000	.40964	.425	.3625	.372	....	.375	000
00	.3648	.38	.3310	.348	....	.34375	00
0	.32486	.34	.3065	.324	....	.3125	0
1	.2893	.3	.2830	.300	.227	.28125	1
2	.25763	.284	.2625	.276	.219	.265625	2
3	.22942	.259	.2437	.252	.212	.25	3
4	.20431	.238	.2253	.232	.207	.234375	4
5	.18194	.22	.2070	.212	.204	.21875	5
6	.16202	.203	.1920	.192	.201	.203125	6
7	.14428	.18	.1770	.176	.199	.1875	7
8	.12849	.165	.1620	.160	.197	.171875	8
9	.11443	.148	.1483	.144	.194	.15625	9
10	.10189	.134	.1350	.128	.191	.140625	10
11	.090742	.12	.1205	.116	.188	.125	11
12	.080808	.109	.1055	.104	.185	.109375	12
13	.071961	.095	.0915	.092	.182	.09375	13
14	.064084	.083	.0800	.080	.180	.078125	14
15	.057068	.072	.0720	.072	.178	.0708125	15
16	.05082	.065	.0625	.064	.175	.0625	16
17	.045257	.058	.0540	.056	.172	.05625	17
18	.040303	.049	.0475	.048	.168	.05	18
19	.03589	.042	.0410	.040	.164	.04375	19
20	.031961	.035	.0348	.036	.161	.0375	20
21	.028462	.032	.03175	.032	.157	.034375	21
22	.025347	.028	.0286	.028	.155	.03125	22
23	.022571	.025	.0258	.024	.153	.028125	23
24	.0201	.022	.0230	.022	.151	.025	24
25	.0179	.02	.0204	.020	.148	.021875	25
26	.01594	.018	.0181	.018	.146	.01875	26
27	.014195	.016	.0173	.0164	.143	.0171875	27
28	.012641	.014	.0162	.0149	.139	.015625	28
29	.011257	.013	.0150	.0136	.134	.0140625	29
30	.010025	.012	.0140	.0124	.127	.0125	30
31	.008928	.01	.0132	.0116	.120	.0109375	31
32	.00795	.009	.0128	.0108	.115	.01015625	32
33	.00708	.008	.0118	.0100	.112	.009375	33
34	.006304	.007	.0104	.0092	.110	.00859375	34
35	.005614	.005	.0095	.0084	.108	.0078125	35
36	.005	.004	.0090	.0076	.106	.00703125	36
37	.004453	.....	.....	.0068	.103	.006640625	37
38	.003965	.....	.....	.0060	.101	.00625	38
39	.003531	.....	.....	.0052	.099	....	39
40	.003144	.....	.....	.0048	.097	....	40

ABBREVIATIONS

METALS

Aluminum	Almn.
Babbitt	Bb.
Brass	B.
Bronze	Bz.
Carbon	Cbn.
Cast brass	C. B.
Cast copper	C. Cop.
Cast Iron	C. I.
Cast steel	C. S.
Cold rolled steel	C. R. S.
Copper	Cop.
Lead	Lead
Malleable iron	M. I.
Open hearth steel	O. H. S.
Phosphor bronze	Ph. Bz.
Steel	Steel.
Steel casting	S. C.
Wrought iron	W. I.
Zinc	Zn.
Tool steel	T. S.
Forged tool steel	F. T. S.
High speed steel	H. S. S.

GAGES

Brown & Sharpe, or American Standard Wire Gage	B. & S.
Birmingham, or Stubs Iron Wire Gage	B. W. G.
National, or Roebling's, or Washburn & Moen's	N. W. G.
Music Wire Gage	M. W. G.
United States Gage	U. S. G.
Twist Drill & Steel Wire Gage	T. D. G.
Stubs' Steel Wire Gage	S. W. G.

FASTENERS

Button head bolt	Btn. Hd. F
Cap screw	Cap Sc.
Double chamfered hexagon nut	Dbl. Chmf
Eye bolt	Eye B.
Fillister head brass machine screw	Fil. Hd. B.
Fillister head iron machine screw	Fil. Hd. I.

REFERENCE TABLES.

Flat head wood screw	Flat Hd. Wd. S
Flat head stove bolt	Flat Hd. Stove
Headless set screw	Hdlss. Set Sc.
Hexagon nut	Hex. Nut.
Lag screw	Lag Sc
Machine bolt	Mach. B.
Machine screw nut	M. Sc. Nut.
Milled body tap bolt	M. B. Tap P.
Set screw	Set Sc.
Square nut	Sq. Nut.
Stud bolt	Stud B.
T-head bolt	T-Hd. B.

WEIGHTS AND MEASURES, ETC.

Center	Cr.
Center line	C. L.
Circumference	Circum.
Diameter	dia. or D.
Foot, feet	Ft. or ', e.g.4'
Horsepower	H.P.
Inch, inches	In. or ", e.g.4"

MISCELLANEOUS

Building	Bldg.
Case harden	C.H.
Company	Co.
Counterbore	Cbr.
Countersink	Csk.
Cylinder	Cyl.
Drawing	Dwg.
General	Gnl.
Hexagon	Hex.
Machine	Mach.
Manufacturing	Mfg.
Maximum	Max.
Minimum	Min.
Specification	Spec.
Square	Sq.
Standard	Std.
Threads	Thds.
Weight	Wgt.
Finish	f.

INDEX

216

www.ingramcontent.com/pod-product-compliance
Lightning Source LLC
LaVergne TN
LVHW012203040326
832903LV00003B/89